Historical Geographies of Prisons

T0188253

This is the first book to provide a comprehensive historical-geographical lens to the development and evolution of correctional institutions as a specific subset of carceral geographies. This book analyzes and critiques global practices of incarceration, regimes of punishment, and their corresponding spaces of "corrections" from the eighteenth to twenty-first centuries. It examines individuals' experiences within various regulatory regimes and spaces of punishment, and offers an interpretation of spaces of incarceration as cultural-historical artifacts. The book also analyzes the spatial-distributional geographies of incarceration, particularly with respect to their historical impact on community political-economic development and local geographies. Contributions within this book examine a range of prison sites and the practices that take place within them to help us understand how regimes of punishment are experienced, and are constructed in different kinds of ways across space and time for very different ends. The overall aim of this book is to help understand the legacies of carceral geographies in the present. The resonances across space and time tell a profound story of social and spatial legacies and, as such, offer important insights into the prison crisis we see in many parts of the world today.

Karen M. Morin is Professor of Geography currently serving as Associate Provost at Bucknell University, Pennsylvania, US.

Dominique Moran is Reader in Human and Carceral Geography, School of Geography, Earth and Environmental Sciences, at the University of Birmingham, UK.

Routledge research in historical geography
Series Edited by
Simon Naylor
School of Geographical and Earth Sciences, University of Glasgow, UK
and
Laura Cameron
Department of Geography, Queen's University, Canada

This series offers a forum for original and innovative research, exploring a wide range of topics encompassed by the sub-discipline of historical geography and cognate fields in the humanities and social sciences. Titles within the series adopt a global geographical scope and historical studies of geographical issues that are grounded in detailed inquiries of primary source materials. The series also supports historiographical and theoretical overviews, and edited collections of essays on historical-geographical themes. This series is aimed at upper-level undergraduates, research students, and academics.

Published

Historical Geographies of Prisons
Unlocking the usable carceral past
Edited by Karen M. Morin and Dominique Moran

Historical Geographies of Prisons

Unlocking the usable carceral past

Edited by
Karen M. Morin and
Dominique Moran

LONDON AND NEW YORK

First published 2015 by Routledge

2 Park Square, Milton Park, Abingdon, Oxfordshire OX14 4RN
52 Vanderbilt Avenue, New York, NY 10017

Routledge is an imprint of the Taylor & Francis Group, an informa business

First issued in paperback 2020

Copyright © 2015 selection and editorial matter, Karen M. Morin and Dominique Moran; individual chapters, the contributors

The right of Karen M. Morin and Dominique Moran to be identified as authors of the editorial matter, and of the individual authors as authors of their contributions, has been asserted by them in accordance with sections 77 and 78 of the Copyright, Designs and Patents Act 1988.

All rights reserved. No part of this book may be reprinted or reproduced or utilised in any form or by any electronic, mechanical, or other means, now known or hereafter invented, including photocopying and recording, or in any information storage or retrieval system, without permission in writing from the publishers.

Notice:
Product or corporate names may be trademarks or registered trademarks, and are used only for identification and explanation without intent to infringe.

British Library Cataloguing in Publication Data
A catalogue record for this book is available from the British Library

Library of Congress Cataloging in Publication Data
Historical geographies of prisons: unlocking the usable carceral past / edited by Karen Morin, Dominique Moran.
  pages cm. – (Routledge research in historical geography)
  1. Prisons–History. 2. Imprisonment–History. 3. Corrections–History.
  4. Human geography. I. Morin, Karen M. II. Moran, Dominique.
  HV8705.H57 2015
  365'.9–dc23   2015000649

ISBN: 978-1-138-85005-7 (hbk)
ISBN: 978-0-367-66877-8 (pbk)

Typeset in Times New Roman
by Wearset Ltd, Boldon, Tyne and Wear

Contents

Figures

Contributors

Clare Anderson is Professor of History at the University of Leicester, UK. Her research has centered on the history of incarceration and penal colonies, and their intersections with other modes of confinement and coerced labor, in the Indian Ocean. Her publications include: *Convicts In The Indian Ocean* (2000), *Legible Bodies* (2004), and *Subaltern Lives* (2012). Her current project – The Carceral Archipelago: Transnational Circulations in Global Perspective, 1415–1960 – is a global history of penal colonies, spanning Europe, Russia, the Americas, Africa, the Indian Ocean, Asia, the Pacific and Australia.

Anne Bonds is Assistant Professor of Geography and Urban Studies at the University of Wisconsin, Milwaukee, US. Her work draws from feminist and anti-racist theories and epistemologies to analyze and contest carceral geographies, poverty, unequal political economic landscapes, and geographies of white supremacy. Her research has been published in *The Annals of the Association of the American Geographers*, *Antipode*, *Social and Cultural Geography*, *Geography Compass*, and *Urban Geography*.

Carrie M. Crockett is a Ph.D. researcher in the Department of History at the University of Leicester, UK. She holds a BA in Russian Language and Literature from Brigham Young University and an MA in English, with an emphasis on Soviet Literature, from the University of Nebraska. Her current research focuses on the development of tsarist penal policy on the island of Sakhalin during the nineteenth century. She has conducted archival research in St. Petersburg, Moscow, and Bishkek, Kyrgyz Republic.

Christian G. De Vito is a member of the Carceral Archipelago team (University of Leicester) and an honorary fellow at the International Institute of Social History, Amsterdam. His current research interests include the social history of confinement and the relationships between "free" and "unfree" labor, with a special stress on convict labor. He is a member of the boards of the International Social History Association newsletter, the International Association "Strikes and Social Conflicts," and the Società Italiana di Storia del Lavoro. He coordinates the working group on "Free and Unfree Labour" of the European Labour History Network and is co-chair of the Labour Network at the European Social Science History Conference.

Susana Draper is Associate Professor of Comparative Literature at Princeton University, US. She is the author of *Ciudad posletrada y tiempos lúmpenes: crítica cultural y nihilismo en la cultura de fin de siglo* (2009) and *Afterlives of Confinement: Spatial Transitions in Postdictatorship Latin American* (2012). She is currently working on another book, *Experiments in Freedom and Cognitive Democracy in Mexico: 1968 OtherWise*.

Katie Hemsworth is a Ph.D. candidate in Geography at Queen's University in Kingston, Canada. Her research interests include geographies of sound, incarceration, emotion, and embodiment. Her dissertation explores historical and contemporary spatialities of sound in Canadian prisons.

Brian Jordan Jefferson is Assistant Professor of Geography and Geographic Information Sciences at the University of Illinois, Urbana-Champaign, US. His research focuses on how criminal justice policies mediate the production and transformation of urban space in US contexts. His work incorporates critical human geography, urban studies, criminology, and social theory.

Carol Medlicott is Associate Professor of Geography at Northern Kentucky University, US. As a historical geographer, she has been researching the Shaker West since 2004. Her work on Shakers has been published in *Timeline of the Ohio Historical Society*, *American Communal Studies Quarterly*, and *Communal Societies*, as well as in several book chapters and monographs, including *Issachar Bates: A Shaker's Journey* (2013).

Takashi Miyamoto is a Junior Fellow at the Research Institute for Languages and Cultures of Asia and Africa, Tokyo University of Foreign Studies, Japan. His study centers on institutional changes of punishment in Japan, India, and the Malay Peninsula during the nineteenth and twentieth centuries. He is especially interested in the international circulation of information on penology. In order to observe how prison officers selected information from a vast information network that was being created throughout the world, he is also exploring a methodology of digital history.

Dominique Moran is Reader in Human and Carceral Geography in the School of Geography, Earth and Environmental Sciences, University of Birmingham, UK. She is the author of *Carceral Geography: Spaces and Practices of Incarceration* (2015), and she has also published from research in carceral environments, both within human geography (e.g., *Transactions of the Institute of British Geographers, Environment and Planning D: Society and Space*) and in criminology and prison sociology (e.g., *Punishment and Society* and *Theoretical Criminology*). She is currently conducting interdisciplinary research funded by the UK. Economic and Social Research Council into the relationship between prison visitation and recidivism, and the philosophy of prison design.

Karen M. Morin's works in feminist historical geography include *Frontiers of Femininity: A New Historical Geography of the Nineteenth-century American West* (2008) and *Civic Discipline: Geography in America, 1860–1890* (2011). She has recently published works on historical geographies of prisons in *Environment and Planning D: Society and Space*; *Men and Masculinities*; and *The Geographical Journal*; as well as in *Critical Animal Geographies: Power, Intersections, and Hierarchies in a Multi-species World* (2015).

Kellie Moss completed her MA in History at the University of Leicester, UK. in 2012, and was subsequently awarded a Graduate Teaching Position with Leicester's School of History in 2013. Her research centers on the global integrations of convicts in Western Australia between the years 1850 and 1868, addressing the circulation and utilization of convicts from within the British Empire. She is also currently an affiliated researcher on the ERC Carceral Archipelago project (2013–2018) coordinated by Professor Clare Anderson.

Jack Norton is a doctoral student in the Earth and Environmental Sciences Department at the CUNY Graduate Center, US, writing a dissertation on the geography and political economy of mass incarceration in New York State during the last forty years. He is a geographer interested in questions of race, labor, development, gender, agrarian change, and political ecology, and is also a research assistant at Hunter College on the NSF-funded *Mapping the Solidarity Economy* project. He received a Master's degree in Geography from the University of Washington, was a Canadian Studies Fellow at the Jackson School of International Studies, and has worked as a researcher and activist in the United States and Canada for over a decade.

Kimberley Peters is Lecturer in Human Geography at Aberystwyth University, UK. Her research focuses on the intersections between place, mobility, and material culture, focusing on these in the context of tourism studies and also geographies of the sea. Kimberley's work has been published in journals including *Area, Mobilities, Tourism Geographies* and *Environment and Planning A*. She is co-editor of the book *Water Worlds: Human Geographies of the Ocean* (2014).

Justin Piché is Assistant Professor in the Department of Criminology at the University of Ottawa, Canada, and Co-managing Editor of the *Journal of Prisoners on Prisons* (2008–2014). He is the 2012 recipient of the Aurora Prize, Social Sciences and Humanities Research Council of Canada. His current research with Kevin Walby examines the cultural representations of incarceration and punishment in Canadian popular culture. Through an analysis of penal history museum narratives, relics, spatial arrangement, and conservation practices, they seek to understand how historical sites contribute to our individual and collective understandings of prison life. He also conducts research on prison expansion and available policy alternatives.

Katherine Roscoe is a Ph.D. student at the University of Leicester, UK, and a member of the Carceral Archipelago team. Her thesis looks at the spatial history of Australian convict islands, both in terms of their natural geography and as built spaces. She completed her Masters in World History at King's College London in 2013 with a project that explored the experience of time in Pentonville Prison in the first half of the nineteenth century. She is the convener of the Carceral Archipelago reading group and co-organizer of the Imperial and Global History Network's Inaugural Conference at the University of Exeter.

Minako Sakata is Associate Professor at Tomakomai Komazawa University. She received a doctorate in Area Studies from the University of Tokyo in 2007. Her special field is ethnohistory and oral tradition studies of the Ainu. In recent years she has engaged in the modern history of Hokkaido, focusing on convict transportation, settler colonialism, and internal colonialism. Her current interests also include comparative oral tradition studies among indigenous peoples in the Arctic Circle, especially on representation of others or contact narratives about colonizers.

Rashad Shabazz is a human geographer of race, gender, and cultural production at the Department of Geography, University of Vermont, US. His forthcoming book, *Spatializing Blackness* (University of Illinois Press), examines how the history of carceral power within Black Chicago informed masculinity, cultural production, and health. His scholarship has appeared in *ACME*; *Souls*; *Gender, Place and Culture*; *The Spatial-Justice Journal*; and he has published several book chapters and book reviews.

Brett Story is a Ph.D. candidate in Geography, and Junior Fellow at the Centre for Criminology and Sociolegal Studies at the University of Toronto, Canada. Her dissertation research focuses on the relationship between solitary confinement, mass incarceration, and urban restructuring in the US over the past 40 years. She holds honors degrees from McGill University and the University of Oxford. She also works as a non-fiction filmmaker and writer. Alongside her dissertation research, she is currently pursuing a film project about the influence of prisons on outside spaces throughout the US, tentatively titled *The Prison in Twelve Landscapes.*

Jennifer Turner is a post-doctoral researcher in the Department of Criminology at the University of Leicester, UK. Trained as a geographer, her current research focuses upon prison architecture and design, and its relationship to penal purpose. Her work has been published in journals including *Environment and Planning A*; *Area*; *Geography Compass*; and *Space and Polity.*

Kevin Walby is Assistant Professor of Criminal Justice at the University of Winnipeg, Canada. He is author of *Touching Encounters: Sex, Work, and Male-for-Male Internet Escorting* (2012). He is co-editor of *Emotions Matter: a Relational Approach to Emotions* (2012), *Brokering Access:*

Power, Politics, and Freedom of Information Process in Canada (2012), *Policing Cities: Urban Securitization and Regulation* (2013), and *Corporate Security in the 21st Century: Theory and Practice in International Perspective* (2014). He is also Co-managing Editor of the *Journal of Prisoners on Prisons*.

Acknowledgments

The idea for the works collected here developed from Karen Morin's Distinguished Historical Geography lecture delivered at the 2013 Los Angeles meeting of the Association of American Geographers. She would like to thank the Historical Geography Specialty Group (especially Maria Lane, Bob Wilson, and Garth Myers) for that opportunity. We would like to thank our authors for participating in both the subsequent sessions on carceral historical geographies at the Tampa AAG in 2014, and the 2015 International Conference of Historical Geographers in London. These have been great opportunities to create new connections among carceral and historical geographers. Sincerest thanks go to Faye Leerink, Emma Chappell, and the whole Routledge team for seeing this project through so professionally in such a short time frame; and thanks as well to editors Simon Naylor and Laura Cameron for launching the new Routledge Series in Historical Geography with this work. Finally we appreciate Bucknell University's valuable financial backing of the project.

1 Introduction

Historical geographies of prisons: unlocking the usable carceral past

Karen M. Morin and Dominique Moran

Introduction

Over the past several decades historical geographers have become increasingly attuned to the fact that they play an active, participatory role in creating the past, and that the pasts they create have tremendous implications for contemporary audiences and places (Schein, 2006, 2011; Nash and Graham, 2000; Alderman and Modlin, 2014; Morin, 2013b). With this recognition comes awareness that there really is no point in studying the past unless there is something we can learn from it. The past must be made relevant, have purpose, and make a difference. To that end this volume brings together the works of self-identifying historical geographers, as well as other scholars who use historical-geographical logics and perspectives (e.g., after Gilmore, 2007), to examine, analyze, and critique practices of incarceration, regimes of punishment, and their corresponding spaces of "correctional" institutions (prisons and jails), with the overall aim of helping to understand their legacies in the present. The breadth of the work collected here spans the eighteenth through twenty-first centuries, and takes a correspondingly wide geographical reach across sites in North America, Europe, Asia, and Latin America. The resonances across space and time tell a profound story of social and spatial legacies and, as such, offer important insights into the prison crisis we see in many parts of the world today (most notably the US; see Alexander, 2012). The project of this book, simply stated, is that in order to understand that crisis we must first pragmatically distinguish a "usable" historical geography of the carceral past (after Tosh, 2008; Blake, 1999; Olick, 2007).

This is the first volume of its kind to take a comprehensive historical-geographical approach to the study of correctional institutions as a specific subset of the new, fast-developing subfield of carceral geography; although the work is, at the same time, closely engaged with a number of related subfields in geography – cultural, economic, political, and urban geography – as well as with a number of other fields of prison research including criminology, prison sociology, and tourism studies. We, for example, examine political-economic settings within and beyond the prison, as well as the social and cultural products and effects of prisons.

Moran (2013) has defined carceral geography as a field of geographical research that focuses on practices of incarceration, viewing "carceral space" broadly as a type of institution whose distributional geographies, and geographies of internal and external social and spatial relations, should be explored (also see Philo, 2012). She observes that whereas scholars from other disciplines have tended to focus on *time* as the basic structuring dimension of prison life ("doing time," etc.), geographers are particularly well positioned to foreground the experience and study of prison space or, alternatively, time-space (Moran, 2013: 175–176). Carceral geographies have, to date, been understood primarily in dialogue with the broad-ranging works of theorists such as Michael Foucault (1979) on the development of the prison and the regulation of space and discipline of the body; Erving Goffman (1961) on the "total institution"; and Giorgio Agamben (1998) on spaces of exception, where sovereign power suspends the law, producing a zone of abandonment. Along with these, our project also aligns with the aspirations of Loyd *et al.* (2012) and Moran *et al.* (2013), who, however, interpret contemporary carceral geographies within a framework that encompasses the broader field of concentration camps, immigration detention centers, and other spaces of confinement beyond correctional institutions.

Although much of the carceral landscape is hidden from view, secreted away, invisible in out-of-the-way rural areas (Glasmeier and Farrigan, 2007; Bonds, 2006, 2009), once we start paying attention, "carceral space," just in terms of the corrections industry alone, seems to be just about everywhere (Martin and Mitchelson, 2009). A whole host of carceral sites and scenes rooted in place – physical structures and buildings as well as their representations – are common, ubiquitous components of everyday life, at various scales. Constructing an historical geographical usable past thus offers rich opportunities for the study of sites of experience and memory in the material landscape at various scales: from the study of spaces and architectures of tourist and heritage sites; to the study of the spatial logics and legacies inherent in the development of prison "archipelagos," urban spaces, and prison towns, particularly along racial and ethnic topographies; to the study of the retrofitting or decommissioning of sites and their re-commissioning for other uses, among others. Such are material carceral artifacts but also "discourse materialized": they tell stories about their contents, and in turn are experienced and understood in a wide range of ways.

Although there are a number of overlapping themes, authors of this volume focus on the historical-geographical study of spaces and sites of corrections in three ways: (1) by examining individuals' experiences within various regulatory regimes and spaces of punishment; (2) by interpreting spaces of incarceration as cultural-historical artifacts; and (3) by analyzing the spatial-distributional geographies of incarceration, particularly with respect to their historical impact on community political-economic development and local geographies.

To date, those who may self-identify as working within the subdiscipline of historical geography to study prison spaces have primarily contributed to a fairly narrow concept of it. For instance, Driver (1985), Philo (1992, 2001), and

Ogborn (1995) studied penal institutions with respect to their interior design and architecture (focusing in particular on Foucault's analysis of Bentham's panopticon), as well as the myriad social practices and tactics internal to prisons used to control ("discipline") people and their movements. Building on this work, contributors to this volume examine a range of prison sites and the practices that take place within them (e.g., sound control, solitary confinement) helping us to understand how regimes of punishment are experienced, and are constructed in different kinds of ways across space and time for very different ends. We might assume, for example, that solitary confinement denotes a space of silence and contemplation as part of the disciplinary regime, such as that associated with many of the early eighteenth-century American prisons (Bruggeman, 2012; see Story, Chapter 3, this volume). But spaces of solitary confinement can be teeming with noise – generated from hygiene fixtures, from the clanging of heavy steel doors, from guards and inmates yelling and shouting. Hemsworth, for example (Chapter 2, this volume), considers the embodied experience of imprisonment by paying attention to such historic "acoustemologies" as important parts of the disciplinary regime, disciplining the body in particular kinds of ways and indeed weighing heavily on it. Carceral spaces also work through local corrections to maintain racial, ethnic, and gender boundaries and identities that in turn become sites of active resistance (Shabazz, Chapter 4, this volume).

How we understand regime shifts in the material carceral landscape and culture is one of the most important questions which scholars struggle to understand (Gottschalk, 2006; Thompson, 2010; Wacquant, 2001, 2009; Peck, 2003; Morin, 2013a). Many correctional facilities have undergone tremendous structural and ideological shifts over their life cycle, and as such may contain "living memories" that are incorporated into their present form. Penal zones and villages within the former Soviet Gulag, for example, tell important stories about integration of the carceral in cultural landscapes of the past and thus impact the way the past is remembered in the present (Pallot *et al.*, 2010). Bogumił *et al.* (2015) have shown these to be co-opted by the Russian Orthodox Church as "sacred spaces" that honor Gulag martyrs, which meanwhile diffuses and even bypasses or erases state responsibility for the horrors of those institutions.

The notion that a prison incorporates living memory in both its physical structure and philosophies of punishment is an important one. There are many what may be considered "historic" prisons or jails, such as those that have been turned into commercially driven tourist sites based on their architectural significance, with local preservation or historical societies supporting their protection (Schrift, 2004; Bruggeman, 2012; Walby and Piché, 2011). Contributors to this volume examine the development of such facilities through their life cycle, collectively arguing that it is important to understand how the carceral state apparatus encourages us to memorialize prison space in particular kinds of ways, as well as understand its impacts on land use and urban space (see Walby and Piché, Chapter 6, this volume). We might consider these as part of the "post-prison" cultural landscape: heritage sites no longer functioning as spaces of incarceration, but nevertheless "still saturated with, and arguably communicative of, messages about the purpose of imprisonment both in terms of the system during which it was constructed, and during which it was protected, conserved, demolished, or left to decay" (Moran, 2015: 4).

Many prisons have become "dark" museum tourist sites, which raise questions about how various feelings of incarceration are orchestrated for the visitor but which the tourist also necessarily co-produces (Strange and Kempa, 2003). How are the "bricks and mortar" – the materiality of the building – combined with the haptic, affective, and emotional spheres to create the carceral experience for specific ends? Is the visitor experience intended to be fun and sanitized, or haunting and sobering? How are the material and the discursive resolved for the visitor? For example, many prison heritage sites suggest that contemporary practices of imprisonment are more "civilized" when compared with the barbarous past, which simultaneously creates a social distance between "us" and "them," the visitor and prisoner. Thus, although the commodification of the macabre at prison sites erases as much as it reveals in relation to the communication of meaning and purpose of punishment to visiting audiences, it also offers opportunities to confront contemporary issues of imprisonment (Walby and Piché, 2011; Moran, 2015; see Turner and Peters, Chapter 5, this volume).

Contributors to this volume consider both how some correctional institutions are re-commissioned as commercial space or sited for other myriad uses; or alternately, how non-prison sites can become repurposed for the correctional industry. Simply as pieces of physical infrastructure, large, sturdy decommissioned prison buildings, if located in commercially viable places, have various potentially profitable uses, such as conversion into housing or hotels, as in the instance of the Hotel Lloyd in Amsterdam (Ong *et al.*, 2012). Such works ask us to question whether prisons can, in fact, ever "be" something else; or indeed whether they should be, considering the potential erasure of past horrors that such sanitizing practices represent (see Draper, Chapter 8, this volume). Thus it is debatable whether a "post-prison" existence is even possible, in which a post-functional prison building takes on a new usage completely disconnected from its previous function. In such cases, even if the former function of the site is incidental to its commercial appeal, arguably the post-prison still remains a prison, and its identity becomes part of the narrative of the re-purposed site – such as the Puna Carretas prison in Uruguay which is now a shopping mall. Medlicott (Chapter 7, this volume) also offers an interesting example of how non-prison buildings and sites such as those of early American intentional religious communities were re-purposed as correctional facilities and are still in use today.

Where prisons are located today is a product of past spatial logics and priorities, which were often under very different social, economic, and political circumstances (Fraser, 2000). Local authorities lobbying for prisons argue that they will bring economic development and jobs (Bonds, 2006, 2009; Che, 2005; Glasmeier and Farrigan, 2007). Moreover, the spatial legacy and spatial fixity of correctional institutions helps create local geographies with multiple and diverse spin-out effects for residents as well as for prisoners and their families who typically are from long distances away. These locational issues of the past have so much present relevance, and are very hard to undo. Spatial fixity and appropriate infrastructure are necessary for the hyper-incarceration we see in many places, and yet such requirements often come into tension with local and regional organizing for other cultural and economic outcomes.

Correctional facilities may be quintessential carceral landscapes, but just as prisons are porous in terms of the movement of inmates, staff, goods and services, communications and so on, they are also porous in that their "transcarceral" nature and techniques pervade the locale, particularly in places which have come to be defined by imprisonment (Dirsuweit, 1999; see Jefferson, Chapter 11, this volume). The phenomenon of prison towns has been observed and studied in the United States, where the prison boom has brought correctional facilities to innumerable (often rural) settlements afflicted by economic disadvantage (e.g., Fraser, 2000). Whether or not rural towns vie for or agitate against prison siting, and whatever the objective impact of prison siting on these rural communities, the trend for new prisons to locate in extra-urban places brings with it a set of controversial implications, outcomes that obscure unfulfilled economic promises, and results that highlight troubling alliances among private and state actors (Moran, 2015; see Bonds, Chapter 12, this volume). The prison boomtown is often part of a larger phenomenon of prison clusterings or "archipelagos" that have been constituted through various geographical scales, mobilities, practices, and historical periodizations. The effect of prison archipelagoes in the US (see Norton, Chapter 10, this volume), again often distributed in rural places abandoned by capital – and in Norton's case, the entire apparatus of the Olympic games in upstate New York – offers important glimpses into the creation and inscription of incarceration on the wider cultural landscape.

Martin and Mitchelson (2009: 461) were among the first geographers to draw attention to the significance of transport, holding facilities, and the mobility of prisoners and guards within and between penal institutions and prison archipelagos. The notion of "disciplined mobility" carries significance with respect to the circulation of ideas, things, practices, and spheres of confinement that has heretofore been neglected within the mobilities literature (but see Gill *et al.*, forthcoming), particularly in how these are personally experienced by incarcerated people. Many of the key questions that have arisen in postcolonial studies resonate with questions about the agency and power geometries of prisoners' everyday lives within these circulations of state administrations and global political economies (Anderson, 2012). Anderson *et al.* (Chapter 9, this volume) offer an intriguing, globe-sweeping example of the impact of moving prisoners long distances in the eighteenth and nineteenth centuries. These historically significant experiences of inmates' coerced or enforced movement, particularly at colonial destinations, greatly influenced the nature and development of the colony as inmates became integrated as part of the citizenry. Such works link emerging historical geographies of incarceration to the more familiar field of historical geographies of colonization, most importantly with respect to the effects on indigenous populations (Jacobs, 2012).

Theoretical context: in search of a usable past

This collection offers a broad range of conceptually rich and empirically grounded scholarship that shares a "usable historical spatiality" as their point of emphasis. Our authors take a variety of historical-geographical approaches to the study of correctional systems and institutions, using what we term, after Tosh (2008: 22–23), a "critical applied historical geography" approach. Underlying the notions of "critical" and "applied" here is our conviction that research should help bring about progressive social transformation. Constructing a "usable" past thus implies taking a pragmatic approach to history and historical geography.

Of course a usable past begs the questions: usable by whom, and for what end or purpose? Borrowing from historians, the term "usable past" has been most often associated with the heyday of social history in the 1970s and 1980s, as a response and challenge to top-down, master narratives and foundational myths primarily surrounding nation-building (Brooks, 1918; Moeller, 1996; Reising, 1986; Rosenzweig and Thelen, 1998). Vigorous debates have ensued over the past several decades among Anglophone historians and historical geographers about the instrumentalism, serviceability, and presentism inherent in attempts to create a usable past; and these from very diverse ideological and methodological positions. Many still hold to the idea that the past is somehow uncritically knowable and accessible on its own terms; others argue that presentist accounts are instrumentally constructed or invented in order to be put to deliberately falsifying, distorting, or manipulative purposes. Questions of reflexivity, positionality, and accountability give us scope to acknowledge that there are different ways to explore the importance of past places, and the importance of the past in place (Schein, 2011: 14–15).

We foundationally acknowledge that historical interpretations are always going to be contingent, since there are always many experiences and perspectives – multiple pasts – happening on the ground simultaneously. Questions regarding the availability of historical sources; archival reason; and archives as sites of power, privilege, and repression, all remain, along with the psychic or psychoanalytic costs of remembering and forgetting traumatic events (at the individual, familial, and social scales). But the usable past as a conceptual framework helps us come to grips with that multiplicity by acknowledging that behind every version of the past are a set of interests in the present; all constructions of the past are in some sense an invention, a "retrospective reconstruction of historical referents to serve the needs of the present" (Olick, 2007: 19). In the case of prisons and jails, we see one pressing "need of the present" to frame the hyper- or mass-incarceration trends of today within well-articulated historical and geographical contexts.

In constructing a usable carceral past it is helpful to turn to historians Rosenzweig and Thelen (1998) who argued that what materially constitutes the usable past includes three elements: (1) the past in particular events; (2) the past as embodied in particular people; and (3) the past as memory materialized on the landscape. Building on this theme, Morin (2013b: 4–6) argues that in constructing a usable *carceral* past the important events and people of that past have already been well integrated into critical prison research and scholarship (e.g., Gottschalk, 2006; Alexander, 2012; Wacquant, 2001; Davis, 2003; James, 2005). But much of this work has been to a large extent space- and place-blind.

Thus first we might note that considerations of carceral space intersect with recent developments in architectural geographies and cultural geographies of buildings (e.g., Kraftl, 2010; Jacobs and Merriman, 2011; Jacobs, 2006; Kraftl and Adey, 2008), where authors have argued for the importance of considering buildings in a number of connected ways. These include the study of buildings as everyday spaces in which people spend a significant portion of their lives; as expressions of political-economic imperatives that code them with "signs, symbols and referents for dominant socio-cultural discourses or moralities" (Kraftl, 2010: 402); and the study of buildings in terms of perspectives that emphasize materiality and affect. This collection takes carceral geography beyond the symbolic meaning of prison buildings to consider their "inhabitation." Like any other buildings, prisons, whether operational or re-commissioned, are sites in which myriad users and things come into contact with one another in numerous complex, planned, spontaneous, and unexpected ways, and where the encounters are both embodied and multi-sensory, and resonant of the power structures which exist both within and outside the prison building and which shape its inhabitation.

Moreover, because most carceral landscapes inherently contain racial, ethnic, and/or class-based biases and uneven power geometries, historical geographies of racialized landscapes also offer useful methodological entry points for study (Schein, 2006, 2009, 2011; Lambert, 2011; Alderman and Modlin, 2014; Alderman and Dobbs, 2011). These resonate with what Franklin (1978: 100), for example, identifies as close to the center of the American historical experience as a nation state, "the plantation to the penitentiary." While most traditional historical geographies of the American "Black experience" have focused on agricultural systems of slavery, migration, urbanization, and labor, more recent trends address segregation, municipal exclusions, consumption practices, and relationships between industrial capitalism and civil rights (Alderman and Modlin, 2014: 273–279; Wilson, 2000). To Alderman and Modlin (2014: 270), the southern plantation museum is "ground zero" in the struggle over "how (or even whether) we should remember past racist practices and landscapes." In that sense such works help us come to grips with how space and geography are deeply embroiled in struggles over racism in the past, present, and future – which in turn influences questions about redress for past wrongs. They also suggest that writing a "usable historical geography" can take an explicit genealogical approach (although it would not necessarily do so). All of the works collected here in some sense adopt a genealogical approach, even if the periodizations under study are perhaps short by standards common in historical research. Schein (2009: 381–383) offers a formula of sorts for the practice of such a genealogy, which involves a four-step process: (1) empirically document when, where, why, and by whom the landscape was created, and how it has been altered over time; (2) ask what the landscape means to the individual and collective identities of people who live in and through it; (3) consider the landscape as facilitator-mediator of particular political, social, economic, and cultural intentions and debates; and (4) ask how the landscape works to normalize/naturalize social and cultural practice, to reproduce those practices, as well as provide a means to challenge those practices. Authors of this volume both implicitly and explicitly follow this formula.

Themes and content of the volume

> I work to tell the truth about people's lives; I work to celebrate struggle, to applaud the tradition of struggle in our community, to bring to center stage all those characters, just ordinary folks on the block, who've been waiting in the wings ... I want to lift up some usable truths.
>
> (Toni Cade Bambara, quoted in Tate, 1983: 18)

Although a variety of topics and approaches could form the scholarly basis for this volume, we limit our purview to three areas: "internal" historical geographies focused on the techniques of incarceration, including insights from prisoner experiences (Part I); the cultural-historical practices and understandings related to the decommissioning or re-commissioning of correctional institutions (Part II); and political-economical analyses of the landscape of prison development, expansion, and contraction (Part III). We acknowledge the many commonalities and overarching points of intersections among the chapters, and that many could be situated in more than one of these subdivisions. However, we have grouped them according to the main points of articulation with the usable carceral pasts.

Part I   On the inside: carceral techniques in historical context

In Part I, **Katie Hemsworth** offers an analysis of the cultural history of sound within Canadian prisons from the early nineteenth century up until today. Her analysis of "sonic mechanisms of power" offers new ways to think about the important role of sound in the way that carceral space is constructed, modified, and remembered. Prisons are multi-sensory, emotional, and embodied experiences for inmates, and, as Hemsworth argues, attempts to control or manipulate sound in prisons tells us much about strategies of control and resistance. **Brett Story** continues the theme of Part I with a chapter on the increasing institutionalization of solitary confinement since the 1960s, analyzing the social versus the individual contradictions of this practice. She argues that inmates with collective access to one another represent social and political threats to the carceral state. Focusing on the social aspect of solitary confinement, Story helps us understand its productive effects and potentialities in challenging the carceral state. **Rashad Shabazz** turns our attention to the "carceral current" that swept through Chicago's Black neighborhoods in the early 1960s. The mechanisms of prison punishment – architectures of enclosure, restrictions on movement and ubiquitous policing, both inside the prison and in neighborhoods, streets, and homes – were the context through which Black gangs emerged, and through performances of Black masculinity created and circulated both inside and outside of prison (Dawley, 1992; Sabo *et al.*, 2001).

Part II Prisons as artifacts in historical-cultural transition

Part II opens with **Jennifer Turner** and **Kimberly Peters'** study of prisons as dark tourist sites that commodify suffering, tragedy, and death for public consumption. Prison museums increasingly use material, performative, and multisensory effects to elicit empathy with former prisoners. Such performances rely on boundary crossings, as visitors experience "spatial dislocation" from the world outside. Drawing on their research at the Galleries of Justice in Nottingham, UK, the authors argue that this is a method by which the carceral past is made usable to contemporary understanding. Next, **Kevin Walby** and **Justin Piché** focus on a process they call "carceral retasking," examining sites in Ontario, Canada which have been turned into dark tourism museums that reproduce imprisonment as a locally significant idea and practice. Their work usefully questions what this means within the context of memorialization and heritage, particularly as it relates to the role of local historical societies which play a key role in re-commissioning sites for contemporary audiences.

Carol Medlicott then examines the transformation of former religious intentional communities in early America into prisons. Focusing on the "Shakers," self-sustaining communities with distinctive architectures and land use, Medlicott shows how their communities' value for compulsory separation and "rehabilitation" for religious or spiritual ends could be repurposed as sites of punishment, particularly during the Progressive Era. Her work offers a number of intriguing questions about the fixity of carceral architectures and landscapes, and the infrastructure necessary for incarceration. Finally, **Susana Draper** outlines an "architecture of memory and affects" that prisoners use to resist practices of repression and confinement. Women from the Buen Pastor prison in Córdoba (Argentina) fought the urban whitewashing of territorial marks of repression (such as escape windows), insisting on an alternative narrative; while those from Santa Martha de Acatitla (Mexico) memorialized their incarceration through works of poetry and mural painting. Draper's work shows how prisoners create historical geographies as a form of collective empowerment that enables them to reflect critically about imprisonment within broader social networks.

Part III Carceral topographies: the political-economy of prison industrial growth and change

Clare Anderson, Carrie M. Crockett, Christian G. De Vito, Takashi Miyamoto, Kellie Moss, Katherine Roscoe, and **Minako Sakata** begin Part III with an examination of processes involved in creating convict destinations and transporting prisoners to these destinations. Using a comparative approach that draws from a broad range of penal colonies and military fortifications across global geographical settings, the authors outline the various legal, administrative, political, cultural, and economic issues that influenced the transportation and incarceration of individuals during the second half of the eighteenth and throughout the nineteenth centuries. **Jack Norton** continues the carceral archipelago theme by examining prison growth and expansion into a cluster of correctional facilities in the Northern New York Adirondack Mountains, an area abandoned by capital in the 1980s and 1990s. Norton focuses in particular on the political and material practices that brought the 1980 Winter Olympics to Lake Placid, which then transitioned into a project of carceral expansion following the games. The intricate politics involved in this transformation illustrate just how wasteful the system is in terms of human life and social capacity for development.

Next, **Brian Jordan Jefferson** turns our attention to another dimension of carceral geography in the late 1980s, to New York City, where the "hyperpolicing" practices of the New York Police Department converted impoverished inner-city zones into intensely regulated, prison-like spaces. Jefferson shows how these practices have overshadowed recourse to incarceration, where New York State's prison population and New York City's jail population have declined precipitously. In so doing he offers new insights into the multiple dimensions of our increasingly transcarceral world. Finally, **Anne Bonds** examines prison privatization trends in the US and controversial alliances and relations among private and state actors. Her chapter focuses on the distinctive political processes that were in play in late twentieth-century Wisconsin which ended in corruption charges against politicians and legislation prohibiting the construction and use of private prisons. Nonetheless, by focusing on the scalar politics involved, Bonds shows how these outcomes had the tendency to reinforce rather than challenge the logics of incarceration.

In the volume's Conclusion **Dominique Moran** draws together the emergent themes from the foregoing chapters, highlighting the value of the explicitly historical-geographical study of correctional systems and institutions, in relation to the notion of a "critical applied historical geography" and the usable past. Foregrounding the notions of "critical" and "applied," this chapter outlines the ways in which this scholarship may be interpreted and deployed in advocating for progressive social transformation, in terms of making the past relevant, purposeful, and usable.

References

Agamben G (1998) *Homo Sacer: Sovereign Power and Bare Life*. Stanford, CA: Stanford University Press.

Alderman DH and Dobbs GR (2011) Geographies of slavery: Of theory, method, and intervention. *Historical Geography* 39: 29–40.

Alderman DH and Modlin EA (2014) The historical geography of racialized landscapes. In Colten C and Buckley G (eds) *North American Odyssey: Historical Geographies for the Twenty-first Century*. Lanham, MD: Rowman and Littlefield, 273–290.

Alexander M (2012) *The New Jim Crow.* New York: The New Press.

Anderson C (2012) *Subaltern Lives: Biographies of Colonialism in the Indian Ocean World, 1790–1920.* Cambridge: Cambridge University Press.

Blake CN (1999) The usable past, the comfortable past, and the civic past: Memory in contemporary America. *Cultural Anthropology* 14: 423–435.

Bogumił Z, Moran D, and Harrowell E (forthcoming 2015) Sacred or secular? "Memorial," the Russian Orthodox Church, and the contested commemoration of Soviet repressions. *Europe-Asia Studies.*

Bonds A (2006) Profit from punishment? The politics of prisons, poverty and neoliberal restructuring in the rural American northwest. *Antipode* 38: 174–177.

Bonds A (2009) Discipline and devolution: Constructions of poverty, race and criminality in the politics of rural prison development. *Antipode* 41: 416–438.

Brooks VW (1918) On creating a usable past. *The Dial* 64 (April 11): 337–341.

Bruggeman SC (2012) Reforming the carceral past: Eastern State Penitentiary and the challenge of the twenty-first-century prison museum. *Radical History Review* 113: 171–186.

Che D (2005) Constructing a prison in the forest: Conflicts over nature, paradise, and identity. *Annals of the Association of American Geographers* 95: 809–831.

Davis A (2003) *Are Prisons Obsolete?* New York: Seven Stories Press.

Dawley D (1992) *A Nation of Lords.* Illinois: Prospect Heights.

Dirsuweit T (1999) Carceral spaces in South Africa: A case study of institutional power, sexuality and transgression in a women's prison. *Geoforum* 30: 71–83.

Driver F (1985) Power, space, and the body. *Environment and Planning D: Society and Space* 3: 425–446.

Foucault M (1979) *Discipline and Punish: The Birth of the Prison.* New York: Vintage.

Franklin B (1978) *The Victim as Criminal and Artist.* New York: Oxford University Press.

Fraser J (2000) An American seduction: Portrait of a prison town. *Michigan Quarterly Review* 39: 775–795.

Gill N, Conlon D, Moran D, and Burridge A (forthcoming) Circuitry and peno-cartography: New directions in carceral geography. *Progress in Human Geography.*

Gilmore RW (2007) *Golden Gulag: Prisons, Surplus, Crisis, and Opposition in Globalizing California.* Berkeley: University of California Press.

Glasmeier AK and Farrigan T (2007) The economic impacts of the prison development boom on persistently poor rural places. *International Regional Science Review* 30: 274–299.

Goffman E (1961) *Asylums.* Garden City, NY: Anchor Books.

Gottschalk M (2006) *The Prison and the Gallows: The Politics of Mass Incarceration in America.* Cambridge: Cambridge University Press.

Jacobs JM (2006) A geography of big things. *Cultural Geographies* 13: 1–27.

Jacobs JM and Merriman P (2011) Practising architectures. *Social and Cultural Geography* 12: 211–222.

Jacobs M (2012) *Assimilation through Incarceration.* Ph.D. Dissertation, Queen's University, Canada.

James J (ed) (2005) *The New Abolitionists: (Neo)Slave Narratives and Contemporary Prison Writings.* Albany: State University of New York Press.

Kraftl P (2010) Geographies of architecture: The multiple lives of buildings. *Geography Compass* 4: 402–415.

Kraftl P and Adey P (2008) Architecture/affect/inhabitation: Geographies of being-in buildings. *Annals of the Association of American Geographers* 98: 213–231.

Lambert D (2011) Afterword: Critical geographies of slavery. *Historical Geography* 39: 174–181.

Loyd J, Mitchelson M, and Burridge A (eds) (2012) *Beyond Walls and Cages: Prisons, Borders, and Global Crisis.* Athens: University of Georgia Press.

Martin LL and Mitchelson ML (2009) Geographies of detention and imprisonment: Inter-rogating spatial practices of confinement, discipline, law, and state power. *Geography Compass* 3(1): 459–477.

Moeller RG (1996) War stories: The search for a usable past in the federal republic of Germany. *American Historical Review* October: 1008–1048.

Moran D (2013) Carceral geography and the spatialities of prison visiting: Visitation, recidivism, and hyperincarceration. *Environment and Planning D: Society and Space* 31: 174–190.

Moran D (2015) *Carceral Geography: Spaces and Practices of Incarceration.* Farnham: Ashgate.

Moran D, Gill N, and Conlon D (eds) (2013) *Carceral Spaces: Mobility and Agency in Imprisonment and Migrant Detention.* Farnham: Ashgate.

Morin KM (2013a) "Security here is not safe": Violence, punishment, and space in the contemporary US penitentiary. *Environment and Planning D: Society and Space* 31: 381–399.

Morin KM (2013b) Distinguished historical geography lecture, 2013: Carceral space and the usable past. *Historical Geography* 41: 1–21.

Nash C and Graham B (2000) Introduction: The making of modern historical geogra-phies. In Graham B and Nash C (eds) *Modern Historical Geographies.* Harlow: Pearson Education Limited, 1–9.

Ogborn M (1995) Discipline, government and law: Separate confinement in the prisons of England and Wales, 1830–1877. *Transactions of the Institute of British Geographers* 20: 295–311.

Olick JK (2007) From usable pasts to the return of the repressed. *The Hedgehog Review* summer: 19–31.

Ong C-E, Minca C, and Sidaway J (2012) The Empire and its hotel: The changing bio-politics of Hotel Lloyd, Amsterdam, The Netherlands. Paper presented at the Royal Geographical Society – Institute of British Geographers Annual Conference, Edin-burgh, July.

Pallot J, Piacentini L, and Moran D (2010) Patriotic discourses in Russia's penal periph-eries: Remembering the Mordovian Gulag. *Europe-Asia Studies* 62(1): 1–33.

Peck J (2003) Geography and public policy: Mapping the penal state. *Progress in Human Geography* 27: 222–232.

Philo C (1992) Foucault's geography. *Environment and Planning D: Society and Space* 10: 137–161.

Philo C (2001) Accumulating populations: Bodies, institutions and space. *International Journal of Population Geography* 7: 473–490.

Philo C (2012) Security of geography/geography of security. *Transactions of the Institute of British Geographers* 37: 1–7.

Reising R (1986) *The Unusable Past: Theory and the Study of American Literature.* New York: Methuen.

Rosenzweig R and Thelen D (1998) *The Presence of the Past: Popular Uses of History in American Life.* New York: Columbia University Press.

Sabo D, Kupers TA, and London W (eds) (2001) *Prison Masculinities.* Philadelphia, PA: Temple University Press.

Schein R (ed) (2006) *Landscape and Race in the United States*. New York: Routledge.

Schein R (2009) A methodological framework for interpreting ordinary landscapes: Lexington Kentucky's Courthouse Square. *The Geographical Review* 99: 377–402.

Schein R (2011) Distinguished historical geography lecture, 2011: Life, liberty, and the pursuit of historical geography. *Historical Geography* 39: 7–28.

Schrift M (2004) The Angola Prison rodeo: Inmate cowboys and institutional tourism. *Ethnology* 43: 331–344.

Strange C and Kempa M (2003) Shades of dark tourism: Alcatraz and Robben Island. *Annals of Tourism Research* 30: 386–405.

Tate C (ed) (1983) *Black Women Writers at Work*. New York: Continuum.

Thompson H (2010) Why mass incarceration matters: Rethinking crisis, decline, and transformation in postwar American history. *Journal of American History* 97: 703–734.

Tosh J (2008) *Why History Matters*. Basingstoke: Palgrave Macmillan.

Wacquant L (2001) Deadly symbiosis: When ghetto and prison meet and mesh. In Garland D (ed) *Mass Imprisonment: Social Causes and Consequences*. London: Sage, 82–120.

Wacquant L (2009) *Punishing the Poor: The Neoliberal Government of Social Insecurity*. Durham, NC: Duke University Press.

Walby K and Piché J (2011) The polysemy of punishment: Dark tourism and Ontario's penal history museums. *Punishment and Society* 13: 451–472.

Wilson B (2000) *America's Johannesburg*. Lanham, MD: Rowman & Littlefield.

Part I

On the inside

Carceral techniques in historical context

2 Carceral acoustemologies

Historical geographies of sound in a Canadian prison

Katie Hemsworth

Introduction

In September 2013, Canada's oldest and most notorious prison, Kingston Penitentiary, closed its doors permanently after 178 years. In a matter of days, it went from being an overwhelmingly noisy institution in Canada to eerily quiet, a ghostly presence on Kingston's waterfront. And yet it is not totally silent. Sound lingers, it escapes, and people escape with it. Inmates and staff carry their sonic histories with them, reshaping these narratives and the new places they occupy. Inside the (not-so-) empty prison, carceral remnants are mobilized by the prison's echoes and resonances: layered emotional histories, thousands of voices silenced over centuries, and the creaks and moans of an utterly exhausted building. There is something hauntingly poetic about an institution that started and ended in so-called silence, and I spend much of this chapter tracing the history of silence as a powerful conceptualization of sound.

One way to engage with a usable past is to consider how the past gets materialized on the landscape (Rosenzweig and Thelen, 1998; Morin, 2013a). I wish to respond to and build from this point with two main claims. The first is that a more robust history of prisons must also involve the sonic materiality of landscape, or what R. Murray Schafer (1977, 1994) terms "soundscape." Following Wener's (2012) work on noise in prisons, I argue that if prisons are experienced aurally, then historical approaches to incarceration should involve asking what prisons have sounded like over the years, and the role of sound in histories of surveillance. Despite the confining and restrictive spatialities of prisons, carceral spaces never exist in isolation. Prisons are thus situated within broader political and cultural contexts, their carceral soundscapes actively reflecting and shaping these contexts.

Second, a sonic appreciation of the usable past necessitates an epistemological approach that is attentive to the spatialities and temporalities of sound. This involves familiarizing ourselves with the material properties of sound and considering how they might give shape to and are constituted by space and time. Intersecting the recent flourishing of carceral geographies (Moran *et al.*, 2012; Moran, 2013a, 2013b; Morin, 2013a, 2013b; see also Bonds, 2006; Martin and Mitchelson, 2009; Philo, 2012), with Steven Feld's (1996) writings on acoustemologies, or sonic knowledges, and Trevor Paglen's (2006) work on carceral recording as method, I argue that "carceral acoustemologies" have much to offer in the development of spatial knowledge about prisons. In constructing an audible past (Sterne, 2003), the methods used by historical geographers should recognize the important role of the ear, and indeed the entire body (Parr, 2010), in the experience of sound and in the retelling of sonic histories.

To illustrate the potential for sound to inform and re-imagine historical geographies of incarceration, I present a Canadian account of incarceration, mainly through the storied past of Kingston Penitentiary in eastern Ontario. Known informally and, indeed, uncomfortably, as the "prison capital of Canada," Kingston has the highest concentration of prisons in the country, having housed ten prisons in the greater Kingston area at various points over the past 180 years. Despite the importance of prisons in the city's collective memory, many prison (counter-) narratives, beyond predictably sensational stories of high-profile criminals, escapes, and riots, remain silenced from Kingston's public discourse. I draw from historical materials collected through Library and Archives Canada, Queen's University Archives (QUA), the Canadian Penitentiary Museum, and Correctional Service of Canada online research archive, to explore the silent system at Kingston Penitentiary from 1835 to 1932 as one of numerous practices that highlights the role of sound and acoustic space in prison operations. Materials consulted include prison regulations and rulebooks; investigative reports and commissions between 1832 and 1934 and in the wake of riots in 1932 (*Report of the Superintendent of Penitentiaries: Kingston Penitentiary Disturbances, 1932*, 1933) and 1971 (*Canada Commission of Inquiry into Certain Disturbances at Kingston Penitentiary during April 1971*, 1973); punishment records between 1835 and 1898; warden and superintendent reports from 1835 to the present; reports of the Office of the Correctional Investigator (see Sapers, 2013); and media accounts of key historic events.[1]

My historical analysis is split into approximately two time periods. The first traces the history and effectiveness of the Penitentiary's silent system from 1835 to 1932, as an example of the disciplinary role of sound in prisons. I then shift to a more contemporary focus, beginning with changes to prison regulations in 1933, following the 1932 riot, and carrying forward to the present day. Here, I outline the evolution of other auditory technologies, including loudspeakers, hidden microphones, and radios as objects that have shaped the material landscape and soundscape of prisons, and which continue to leave their mark on the sites and subjects of incarceration today. Following Trevor Paglen's work on recording carceral landscapes (2006), and Richard Wener's chapter on noise in carceral settings (2012: 189–202), this chapter culminates with a discussion of the methodological promise of sound for histories of incarceration. In the wake of increasing privatization, secrecy, and silence that restricts much research in and about Canadian prisons (Piché, 2012; Yeager, 2008; see also Morin 2013a), sonic research methods – like audio recording and oral interviewing – are subject to great scrutiny and suspicion from correctional authorities. As Paglen (2006: 56) notes, these enactments of authoritative power over sound, even before entering prisons for research, move sound projects "out of a textual regime and into a legal one." Yet for those, like Paglen, who choose to take up the challenge of finding creative circumventions of overbearing restrictions, sonic methods can help permeate institutional silences, opening up critical space through which to explore untold – or retold – narratives of incarceration, and asking new kinds of questions. What did prisons sound like in the past, and what can this tell us about historical relations of power? How have cultural practices of listening and mediating sound in Canadian prisons evolved since 1835? What kinds of sonic methods might aid in the articulation of carceral pasts? A cultural history of sound in prisons is significant not only because it is vastly underdeveloped, but also because it provides an important context for understanding and changing our contemporary approaches to incarceration, justice, and rehabilitation.

Kingston Penitentiary: 1835 to 2013

Kingston, Ontario, a city of approximately 125,000 people, is known as the "Prison Capital of Canada," currently hosting eight prisons of varying security levels located in its surrounding area. Kingston Penitentiary, or "The Pen," is Canada's first and most notorious prison, officially opening in 1835. Initially called "The Provincial Penitentiary of the Province of Upper Canada," it was built on the waterfront of an area historically known as Portsmouth Village. Its main building is a cavernous structure with rows of cells radiating from a central monitoring tower, and capped with a reverberating dome. Although the penitentiary's population has been predominantly male, it housed small populations of women and children in its first 100 years.[2]

Kingston's other nickname, "Limestone City," refers to the abundance of limestone in the area, much of which has been quarried and used to construct prominent city buildings by convict labour from what is now Kingston Penitentiary in the mid-1800s (Hennessy, 1999). Under the silent rule (see below), inmates' voices would rarely be heard in the quarry, yet the sounds of limestone extraction were a key soundmark (Schafer, 1994) of Portsmouth Village, where the penitentiary and the quarry were located. Prisons and prisoners have therefore played an active role in Kingston's construction and its identity as a prison town, though most residents have never stepped through the penitentiary's gates. For visitors, the exterior of the prison is a popular "dark tourism" destination as a notorious site of suffering and punishment (Piché and Walby, 2010; see also Foley and Lennon, 1996).

In September 2013, after 178 years as a federal maximum security institution, Kingston Penitentiary closed, leading to the relocation of approximately 500 inmates to other existing facilities. At this juncture of prison closure and relocation, and amidst troubling practices of hyper-incarceration and overcrowding, it is important to think about what still lingers throughout the soundscape of the abandoned prison, what sits buried in silence, and what we can learn from the penitentiary's audible past in order to change the future of incarceration.

Sonic mechanisms of power

I point frequently to the methodological capacity of sound in this chapter as an undervalued way of knowing, experiencing, and reimagining incarceration. Yet this potential is presented in tension with the place of sound in various embodied "pains" of imprisonment (Sykes, 1958; see also Crewe, 2011; Ignatieff, 1978). Indeed, Sykes' work on carceral suffering and deprivation, later expanded by Foucault (1977), pinpoints a shift from focusing mainly on the infliction of physical pain to *embodied* suffering of various physical, psychological, and emotional oppression in prisons. More recently, sociologists and ethnographers have expanded Sykes' criminological work, including Ben Crewe, whose metaphor of "tightness" (Crewe, 2011), as a conceptualization of suffering, is particularly profound. However, such accounts would benefit from a richer account of carceral sound, silence, and noise. As Wener (2012) has documented, one of the first and most common observations people make upon entering a prison is that it is filled with noise: conversations, footsteps, clanging metal, old ventilation systems, and authoritative voices over a public address system. Focusing primarily on psychological well-being, he argues that this cacophonous soundscape has negative effects on embodied experiences of prisons.

Some of the pains experienced in prison are explicitly sonic (as in the case of overwhelming and unrelenting noise), others are exacerbated by sound (posttraumatic stress), and further pains are alleviated by sonic techniques (headphones). Binary soundscapes of silence and cacophony are stitched together by the discomfort they often elicit: both sonic extremes are extensions of enclosure, confinement, and metaphoric "tightness" that restricts incarcerated bodies through the interplay of firm and soft power (Crewe, 2011). At the same time, sonic tactics have been used by inmates to break through these restrictive walls and alleviate the tightness, as do sonic methods that may offer a critical reevaluation of prisons from beyond their bounded structures.

Modeled after Auburn State Prison in New York, which operated under a strict code of silence and stood in contrast to Philadelphia's segregation model, Kingston Penitentiary's design and penal practices reflect Foucault's (1977) articulation of discipline, punishment, and biopolitics that emerged alongside the penitentiary. Stretching this into a soundworld, Bull and Back (2003: 5) remind us that "the history of surveillance is as much a sound history as a history of vision." If power relations materialize through technologies like the oft-cited panopticon, the "all-seeing" design and theory of surveillance conceived by Jeremy Bentham in the late eighteenth century, the same may be said about the attempts to use sound as a form of discipline and surveillance (Gallagher, 2011), albeit in different manifestations. Although rarely acknowledged, Bentham (2013 [1791]) included in his panopticon sonic design features that were to further satisfy a disciplinary logic: "conversation-tubes" allowed inspectors to give orders to prisoners and to listen into cells without being visible. He explains:

The ... set of conversation-tubes is to enable an Inspector in the Lodge to hold converse in his own person, whenever he thinks proper, with a prisoner in any of the Cells. Fixed tubes, crossing the Annular-Well and continued to so great a length being plainly out of the question, the tubes, for this purpose can be no other than the short ones in common use under the name of *speaking-trumpets.*

This system of tubes, which may be referred to as a 'pan*aud*icon' (for a more contemporary example see Rice, 2003), was abandoned partly because it failed to reproduce the dissymmetry of the panopticon tower, which allowed authorities to watch without being seen by inmates. It is much more difficult to control the movement of sonic stimuli; the tubes worked as a two-way communication technology, which led, at least in part, to its abandonment. However, when used under the common rule of silence, the tubes would still have provided a means of aural control that could purposely disrupt and disorient an unsuspecting inmate. More recently, auditory technologies, such as microphones and loudspeakers, have reintroduced the dissymmetry of aural command and eavesdropping as a form of surveillance sought in earlier prison design, a point to which I will return below.

In 1848, George Brown was commissioned to head an investigation of Kingston Penitentiary amidst increasing allegations of cruel and unusual punishment and poor prison conditions under the rule of Warden Henry Smith.[3] Brown's scathing report (Canada Provincial Penitentiary Commission, 1849; hereafter referred to as the Brown Commission) outlines the prison's "most frightful oppression – revolting inhumanity" and questions the effectiveness of a silent system nearly 15 years into its use. Sound, at the time of the report, was a key problem which prison authorities attempted to control with harsh disciplinary measures (see below).

The notion of silence, as one conceptualization of sound, is a useful approach for exploring historical geographies of prisons, given its historical role as a mode of disciplinary power and form of spatial knowledge. Silence occupies liminal spatial status as simultaneously vast and enclosing (Maitland, 2008). Space is never truly void of sound; even when undetected by human ears, sound-waves constantly engulf and shape space. In prisons, silence has been used and interpreted in a variety of ways (Paglen, 2006; LaBelle, 2010), from reducing human-made sounds, to sensory deprivation, to the systematic oppression and silencing of political voice. As I outline below, recognizing the escaping capacities of sound, an attempt to confine sound was another step to confining prisoners, at least in theory.

Silent treatment at Kingston Penitentiary: 1835 to 1932

Under the principles of Auburn's silent system, prisoners at Kingston Penitentiary were ordered to "preserve unbroken silence" (Upper Canada Legislature, 1836; see also Brown Commission, 1849; Brown, 1850). Initially proposed by

Quakers who sought reform through penitence instead of violence and exile (but who were also the first to denounce it after seeing its effects), the silent system became valued by prison authorities as a tool to re-spatialize and re-temporalize inmates' sense of aurality in an effort to produce docile bodies through "subtle coercion" (Foucault, 1977: 137). Occasionally, subtle coercion required enforcement. As Curtis *et al.* (1985: 28) note, "anything that might disturb the silence and harmony of the institution [was] forbidden under pain of severe corporal punishment." Similar to other material features of an environment that individualize movement and "thin out" or stratify space (see Sennett, 1994), silence was used in the nineteenth-century penitentiary to stratify spaces and the individuals occupying them. Indeed, as outlined in a report which set out plans for the penitentiary, "by this system every prisoner forms a class by himself" (Thomson and Macaulay, 1832), stripping inmates of mutuality and familiarity with their incarcerated peers. In this regard, the Auburn model of silence integrated aspects of the Philadelphia model of segregation through the stratifying properties and uses of sound.

Silence at Kingston Penitentiary was conceptualized as an auditory, spiritual space in which ideological notions of purity, faith, and repentance resided (Thomson and Macaulay, 1832; also see Jackson, 1983; Oliver, 1998; Hennessy, 1999). Bodily sounds were limited to those that rendered audible certain techniques of biopower. Marching feet and the rhythmic sounds of manual labor were acceptable at particular places and times as sonic signals of disciplinary order, while other bodily sounds were deemed unacceptable noise (Curtis *et al.*, 1985; Oliver, 1998) and met with severe punishment. Indeed, punishment records from the nineteenth century (QUA, 1835–1853; 1843–1869; 1850–1856) show a disproportionately high number of instances in which inmates were punished with bread and water diets, cat o' nine tail lashings, and dark cell confinement for breaking the silent rule. Distinctions were made between verbal sounds, other vocal sounds like laughing, crying, or singing, and other bodily sounds such as stomping and clapping (QUA, 1835–1853). Under the silent system, the penitentiary could be characterized as a high-fidelity soundscape – one that contained distinct, easily discernible sound events in a high signal-to-noise ratio – that seemed to amplify otherwise mundane sounds. This is a key difference between the penitentiary's soundscape toward the end of its operation: interview participants identify the prison in its final years as an extremely noisy, low-fidelity environment, featuring a low signal-to-noise ratio with dense layers of sounds. Silence is therefore not simply a conceptualization but also a spatial *structure* of sound, "filling" and constructing carceral space in different ways than noise, marking distinctions between different historic periods.

The success of the silent rule at Kingston Penitentiary depended on strict enforcement by the keepers, known today as guards or correctional officers. Initial rules and regulations outlined by the House of Assembly (Upper Canada Legislature, 1836) stated that keepers "shall require of Convicts, labor, silence, and strict obedience." Punishment records during the penitentiary's first quarter-century of operation offer a glimpse into the struggle to maintain order and

discipline through silence. Children, who were sentenced to Kingston Penitentiary in its formative years, were particularly disadvantaged by the silent rule.
Emotionally charged laughter and audible expressions of fear were punished
with bread-and-water diets, lashings, and dark cell confinement (Brown Commission, 1849; see also Hennessy, 1999). In 1846, for instance, seven children
were lashed with rawhide for "talking and laughing when at dinner" (QUA,
1843–1869). One of these children, Antoine Beauché, who appeared to suffer
from a mental illness, was cited in the Brown Commission (1849: 201) as a frequent target of violent lashings for noisy outbursts, and was often gagged in
attempts to silence him. Here, as well as in punishment records for men and
women (Library and Archives Canada, 1943; QUA, 1850–1856), silence materialized historically as a form of violence, through both the political act of denying
a person their voice and that of physically choking inmates and their ability to
make sound.

The difficulty inmates had in obeying the silent rule, as well as their advant-
ageous use of undetectable sounds to resist confinement, exposes the system of
sonic control as theoretically powerful yet easily betrayed by its practical fragil-
ity. Indeed, as cited in the Brown Commission (1849: 116), witness Reverend
R.V. Rogers testified that:

> the silent system is not at all carried out; the men talk and laugh in groups
> together through the yard, constantly; they know every thing going on
> outside, and the want of discipline is quite notorious and often noticed by
> strangers.

Despite the challenges of enforcing silent behavior, the rule remained in place
from 1835 until the end of 1932 (Canada, *Penitentiary Regulations 1933*, 1934),
after which silence was only demanded at particular times (Dominion of Canada,
Annual Report of the Superintendent of Penitentiaries, 1934: 23):

> By regulation, as amended on January 1, 1933, wardens of penitentiaries are
> permitted to allow convicts to converse in a conversational tone before pro-
> ceeding to work in the morning, during lunch hour, and up to seven o'clock
> in the evening, while confined in their cells.

The silent rule was only informally enforced thereafter. According to the report,
there was little evidence that the abolition of the silent rule threatened security,
though wardens expressed concern that inmates' conversations held no reforma-
tive value. Today, disciplinary silence is used primarily alongside sensory deprivation and segregation in solitary confinement (Guenther, 2013), but it also
persists as an act of symbolic and systematic violence through ex-
communication, the silencing of "voices," and changes to parole restrictions on
community activism (see e.g., *Safe Streets and Communities Act*, 2011).

The fluid, reverberating properties of sound that sound scholars might otherwise champion were – and in many cases still are – understood as threats to

authoritative control and carried ideological underpinnings of disease and impurity. This notion follows Peter Bailey's work on the distinction between sound and noise as something that developed in the nineteenth century with the bourgeois fear of the crowd (1996; see also Attali, 1985). From the outset, unauthorized, prisoner-produced sounds were considered transgressive and even contaminating (Thomson and Macaulay, 1832; Brown Commission, 1849; Evans, 1982). The failure of the silent rule was in large part connected to the uncontainable qualities of sound; if inmates at Kingston Penitentiary could use sound to permeate multiple spaces without moving, the mandate of "separate and away" as a form of decontamination was jeopardized.

Throughout the penitentiary's first century, a docile prisoner was a silent one. The current soundscape of a typical Canadian, cell-style prison exists in stark contrast to the penitentiary in the 1800s, now typically characterized as noisy, except in spaces of solitary confinement.[4] As one participant explained, with the influx in noise and sonic devices, silence has become so rare among the general population that it acts as an alarm signalling danger. Changing technologies have allowed old surveillance systems, like the panopticon, to become entangled with current forms of micro-management; loudspeakers, hidden microphones, and sonic design features, which shape the way sounds travel throughout particular spaces. As I explain below, these transformations continue to leave "soundprints" (Atkinson, 2007) on prisons and their subjects.

Contemporary confinement and technological change: 1933 to the present

Kingston Penitentiary's most infamous riot in 1971 lasted for four days, culminating with the deaths of two inmates and the eventual release of guard hostages. An inquiry led by Justice J.W. Swackhamer (*Canada Commission of Inquiry*, 1973) outlines a list of grievances that precipitated the riot, including: poor living conditions; monotony (notably a word with sonic roots); overcrowding; overguarding; and fears regarding transfer to a new high-tech, maximum-security prison. Amidst the riot's debris lay the prison bell, shattered by the blows of inmates seeking retribution for its years of noisy, unrelenting command. Among archival testimonies the bell is frequently cited as the most symbolic piece of property damaged in the riot and an especially loathed mechanism of control in the prison.

As former inmate, Roger Caron, indicated in his prison memoirs (Caron, 1978; see also Caron, 1985), the interpretations of the bell's sonic command depended on which side of authoritative rule an individual was placed. For the guards the bell was an ally, a technology that saved them energy and promoted the kind of efficiency sought in carceral environments (see also Jewkes, 2002). However, even the guards likely had differing opinions of the bell, depending on spatial and temporal context. In the late nineteenth and early twentieth centuries, guards were required to live within the sound of the prison bell so that they could hear and respond to its audible signals in the event of a riot (Oliver, 1998;

Hennessy, 1999). Highlighting the shifting spatial relations of power, guards were also disciplined sonically, training their ears and bodies to filter auditory signals. The bell links the past prison with the contemporary – a relatively simplistic yet efficient technology that commanded disciplined movement in the 1835 "Pen" as it did leading up to the 1971 riot and onward. The context in which a bell is heard, however, has shifted with broader penal change.

For those who experienced the bell's commanding grip as another form of dehumanization, it is not difficult to appreciate the fury that went into the silencing of the bell. Sounding 32 times daily, it was similar to other tools of torture, both physically because of its loud vibrations, and mentally because it signaled their almost complete lack of control over their own bodies while causing psychological and emotional strife (Crewe, 2011). As Caron recounts (1985: 110), the molded brass object took on a life of its own: a "bullying" monster that barked orders, tortured sleeping bodies, and physically agitated inmates' central nervous systems. Indeed, his descriptions of the bell's destruction read more like the killing of a living beast than an inanimate object, implying that the bell had its own agency to bully and fight their blows. The fact that inmates targeted the bell specifically, rather than those who operated and conceived it, is indicative of its importance as a tool of discipline and symbol of oppression.

Not long after the riot, and alongside broader technological change, Canadian prisons assumed a shift in auditory technologies of spatial and temporal control. Perhaps the most notable related to the electronic public address (PA) system. Used as another form of space–time compression, PA systems in prisons are used to give orders to inmates, both individually and collectively, a key "improvement" on the bell. In his work on the loudspeaker, Seth Cluett (2010: 9) writes of the loudspeaker as "[enabling] a return to orality that extended language into space in the same way that literacy helped extend writing over time" (see also Cluett, 2013; Ong, 2002). As a sonic technology that has been manipulated to be executed at high decibel levels and to penetrate virtually every space in the prison, inmates and staff have little agency to ignore it. Instead, their bodies become tethered to the robotic sound of the loudspeaker, and are forced to constantly "keep an ear out" for orders from authority figures. Perhaps most importantly, the PA system enforces and agitates daily rhythms, at times punctuating otherwise steady flows of work and whatever forms of silence or solitude a person desires in an otherwise cacophonous carceral soundscape.

As Wener (2012) warns, this type of noise can have non-acoustic impacts, such as sleep deprivation, stress induction, agitation, and reduced concentration, many of which exacerbate or lead to depression. Relative continuity and calmness are disrupted by one voice demanding the attention of many, constituting the PA's "sonic dictatorship" as noise that cannot be silenced or controlled by constrained inmates. The sonic quality of a loudspeaker varies considerably from the clanging, resonating effects of a bell, typically producing grainy, mechanized sounds that physically interact with bodies and leave their marks on the carceral soundscape.

The bell and loudspeaker are technologies that project sound outward; sound *capturing* devices have also evolved to heighten auditory surveillance in prisons.

Gone are the plans of Bentham's "listening tubes," but technological innovations like the microphone have reintroduced the notion of asymmetrical surveillance. Now small enough to be hidden throughout spaces of visitation as a way to discretely record conversations, microphones facilitate "eavesdropping" on inmates, their visitors, and even staff members. Signs left on the walls of Kingston Penitentiary following its closure alerted people that they may be recorded at any time, encouraged self-monitoring and obedience. This practice of eavesdropping calls for pause, as it highlights the potentially oppressive nature of aurality (Sterne, 2003; Goodman, 2010) and encourages us to question the ethics of listening and of recording sound (see also Paglen, 2006). As researchers, we must also remember that sonic methods for engaging with past and present carceral spaces are never removed from this politics of aurality.

Revisions to penitentiary regulations in 1933 following the 1932 riot saw changes to leisure affordances (Dominion of Canada, 1933; Marr, 1988). The resulting slow introduction of radios into Canadian prisons was highly transformative in the reconstruction of carceral soundscapes. After the silent rule, the ability to listen to music and other radio programs was novel, with radios (and later televisions) quickly becoming a "cherished opiate of the convict" (Hennessy, 1999). Although still present and similarly cherished today, the radio had contradictory uses and responses. It was effective as a community-building tool, given that it provided opportunities for communal listening and fostering an awareness of worldly issues, but these benefits were contradicted by its propensity to produce conflict. This was particularly evident as the portability of the radio turned listening into a more individual experience in each cell. Before headphones became widely available, many of these conflicts originated over the noisy "invasion" of one's personal acoustic space by another's radio, adding to already overbearing restriction and confinement (Crewe, 2011). Even recently, through the telling of Kingston Penitentiary's past upon its closure, visitors' anecdotal stories whisper that the last person to be murdered by a fellow inmate was killed over a dispute about radio volume. The emergence of individual radios in the 1950s to 1960s, partly through the formation of the Correctional Planning Committee in 1960 (Correctional Service of Canada, 2010; Ricciardelli *et al.*, 2014), offered a greater sense of personal agency because inmates could control, to some extent, listening options and locations. Unsurprisingly, this sense of control remained relative. Despite predictable conflicts about volume levels, it was not difficult to see that these "aural pacifiers" could further aid in the production of docile, incarcerated bodies.

The radio served as a way to broadcast recorded voices to a wider public, breaking down and through some of the walls that previously severed communication and facilitated the rejection of inmates from surrounding communities (Sykes, 1958). Indeed, inmates who operated and participated in Kingston Penitentiary's short-lived radio program from 1952 to 1955, *Kingston Penitentiary is On the Air*, contributed to a special audible history that has shaped the penitentiary as we know it today, but that may have been forgotten in the public remembering of its past. Regrettably, under the current regime in Canada, prison programs such as radio shows and music performances, which once offered aural expression and connection to outside communities, have been discontinued.[5]

The current conditions of incarceration in Canada reflect a crisis of double-bunking and overcrowding (Sapers, 2013; see also Ricciardelli *et al.*, 2014). Given that incarcerated populations fluctuate with policy change (see Pizarro *et al.*, 2006; Ricciardelli *et al.*, 2014), this latest trend of overcrowding can most recently be tied to the *Safe Streets and Communities Act* (2011), an omnibus bill that introduced harsher sentences, new criminal offenses, and mandatory sentencing. In other words, "people will be incarcerated for longer periods of time and more frequently" (Ricciardelli *et al.*, 2014). The resulting overcrowding will have an impact on sonic experiences of incarceration. Crowded conditions are noisy. They act as primary triggers of pain (Sykes, 1958) and deepen oppression in contemporary Canadian prisons, increasing desperation to reclaim personal space. As Wener (2012: 197) indicates, crowding heightens noise levels, since "much of the noise in correctional settings is generated by the people there." In addition to the very serious physical impacts of noise in prisons, including nausea and headaches (Stansfeld and Matheson, 2003) and possible hearing loss (Jacobson *et al.*, 1989), constant exposure to noise has myriad emotional and psychological effects (Wener, 2012). A loss of privacy is in part experienced as a sonic threat to dignity and autonomy (see Rice, 2013), as multi-layered noise infiltrates people's sense of personal acoustic space.

The introduction of headphones in prisons in the past two or three decades offered some relief in the mediation of carceral soundscapes. In the past few decades, headphones have become prized commodities that help reclaim personal acoustic space while also filtering out undesirable sound. Similar to personal radios, but more effective in the hushing of auditory media, headphones blur the boundaries between agency and powerlessness. The ability to create a private soundworld while filtering out unwanted noise offers a small, but crucial, opportunity to reclaim control and autonomy that is otherwise deprived upon incarceration. However, a key concern remains: prison authorities may likely use headphones as pacifying technologies for the masses, instead of addressing the more threatening problem of overcrowding directly.

Carceral acoustemologies: sonic approaches to "doing time" (and space)

In this section, I borrow from Dominique Moran's (2012) carceral play on words to explain how scholars might use sound and sonic methods to understand "doing" time: that is, by exploring the methodological potential of sound to articulate carceral time as well as space. Cautioning against the privileging of vision – or "ocularcentrism" – in geographical knowledge, Gallagher and Prior (2014) challenge the assumption that "earwitnessing" is less reliable than eye-witnessing (see also Smith, 1997; Sui, 2000).

Sound is a useful conceptual tool for historical geography as something that is at once tangible and intangible, material as well as discursive. After all, we *feel* sound: we are physically and emotionally moved by vibrations, and we also use sound to construct space and time. A "carceral acoustemology" is indeed a useful conceptual tool, but its us*ability* is perhaps more challenging to negotiate, not least because of the political climates in which researchers and prisons are situated. Oral histories are hard to obtain when requests to engage deeply with inmates about personal experiences of sound are increasingly blocked by correctional authorities (see Piché, 2011), as are audio recordings that are disallowed in the name of safety and privacy (Paglen, 2006). Beyond the challenges presented by heightened restrictions in both prisons and archives, geographical research that takes an historical approach would benefit from more extensive training on aural methods (Gallagher and Prior, 2014) that are reflexive of the productive and disruptive uses of sonic technologies.

Approaching the usable past with a sonic sensibility may not always send us on a new search for historic auditory materials, nor does it necessarily require geographers to become experts in phonographic methods. As a starting point, we might simply revisit material remnants of the past with a redirected focus – to be *attuned* to that which we have overlooked – rendering something newly informative. Audio recordings of speeches inside prison walls, for instance, have useful content beyond the words spoken; "voice" is not simply verbal content but may also be analyzed for aurality and soundscape (Kanngieser, 2012; Gallagher and Prior, 2014). Ambient sound in existing recordings (echoes, resonances, audio devices) may say something about the acoustic characteristics of a transformed space, or is largely unexperienced by the non-incarcerated public. Similarly, visual evidence, such as photographs of past and present architecture, events, or surveillance mechanisms (see James, 2014), may be revisited with an imaginative ear, eliciting an affective atmosphere (Anderson, 2009) that adds nuance to artistic representations.

Other emergent sonic methods offer rich potential for capturing the spatial characteristics of a given site, such that the essence of a past soundscape can be reconstructed long after it has evolved. Historical geographers may begin with soundwalks: aural exercises in deep, mobile listening (see Schafer, 1977, 1944; Westerkamp, 1974). Adapting the soundwalk method to one that is more historically layered and self-directed, Michael Gallagher (2014) has introduced the "audio drift," stitching together stories, music, and ambient sound clips to capture the evolving history and atmosphere of ruins, effectively intersecting past and present. Similarly, impulse response (IR) tests (Cabot, 1978) offer an indication of the "acoustic signature" of a cell in a particular time and place and exemplify one way to simultaneously capture and reconstruct past soundscapes. This may be particularly useful to show differences in prison soundscapes as they vary by time and space, to recreate a listening experience that has been lost to changes in prison design.[6] Such projects need not only be undertaken by researchers using participatory historical methods (see Cameron, 2014); inmates at a minimum-security prison near Kingston, for example, acted as historians and carceral commentators through their collaborative production of a music album, *Postcards from the County* (Brown, 2014), which interweaves personal narratives and soundscapes of past, present, and future.

As is evident in the above examples, sound-based methods encourage experimentation with acoustic space as a way of *in*voking and *e*voking carceral pasts. In doing so, these methods enable rich collaboration and may foster empathy through an emphasis on the politics of listening. Echoing the work of Piché and Walby (2010), however, I caution against researchers using such methods in prisons as another manifestation of carceral tours, which can lead to degrading treatment of inmates. Instead, and always reflexively, incarcerated populations may be engaged in facilitated listening and sound-making exercises, in which they create their own narratives of carceral space, becoming more active in constructing a history that might otherwise be silenced simply because of their incarceration.

Tracing histories of sound and sonic control in prisons through Kingston Penitentiary's silent system and evolving auditory technologies, while making a case for an "acoustemological" approach, I have urged geographers to take up cultural histories of sound in their explorations of usable carceral pasts. Material reminders of past soundscapes and sonic relations of power linger among us with the potential to be creatively revived. We might capture, remember, or speak to this past by re-entering prison sites and experimenting with the sonicity of these spaces, or by re-engaging carceral histories with a revived sonic imagination. Methodologies that embrace the combined material and discursive properties of sound can help to weave the contemporary with the historic, the noise with the silence, and the presences with the absences, as we endeavor to change the face of incarceration as we currently know it.

Notes

1 The broader research project also includes interviews conducted between 2012 and 2014 with people who have worked and lived inside Kingston-area prisons. Although their accounts are paramount for understanding personal experiences of incarceration and are incorporated into other writings, I only draw upon them occasionally in this chapter to help contextualize contemporary prison soundscapes.
2 Women were housed in Kingston Penitentiary until 1934 and were most often punished with bread-and-water diets in dark cells for laughing, singing, and talking (QUA, 1850–1856). I do not have the space here to discuss the shift from Kingston Penitentiary as a mixed-gender to a male-only institution; however, my broader project explores differences in pitch levels, punishment, types of work, leisure, and sonic events like childbirth as a way of exploring how carceral soundscapes become gendered.
3 The tone of the Brown Commission seems warranted given the alarming prison conditions; however, it should be noted that Warden Smith was a close friend of Canada's first prime minister, Sir John A. Macdonald, who did not take kindly to Brown's use of *The Globe* as an outlet for criticizing the government (Oliver, 1998).
4 For a fuller account of solitary confinement ("administrative segregation") in Canada, see Brett Story (Chapter 3, this volume).
5 However, CFRC 101.9, Kingston's community and campus radio station, airs a weekly program called *CFRC Prison Radio (CPR)*, whose activism-based programming features prison justice issues and "Calls From Home," a segment in which inmates can communicate with community members through song requests and messages.

6 Further, audio files from IR tests may be employed to transfuse a carceral soundscape into a non-carceral setting. An example of this is demonstrated by sound artist Matt Rogalsky, whose production of rock band PS I Love You's album *For Those Who Stay* (2014) included impulse response-based reverbs recreating the acoustics of various locations within Kingston Penitentiary.

References

Anderson B (2009) Affective atmospheres. *Emotion, Space and Society* 2(2): 77–81.
Atkinson R (2007) Ecology of sound: The sonic order of urban space. *Urban Studies* 44(10): 1905–1917.
Attali J (1985) *Noise: The Political Economy of Music.* Minneapolis: University of Minnesota Press.
Bailey P (1996) Breaking the sound barrier: A historian listens to noise. *Body & Society* 2(2): 49–66.
Bentham J (2013 [1791]) *Panopticon: or, the inspection-house … In a series of letters, written in the Year 1787, From Crecheff in White Russia, to a Friend in England. By Jeremy Bentham, of Lincoln's Inn, Esq. Dublin, M.DCC.XCI.* Eighteenth Century Collections Online. Gale Group.
Bonds A (2006) Punishment and profit? The politics of prisons, poverty, and neoliberal restrictions in the Rural American Northwest. *Antipode: A Radical Journal of Geography* 38(1): 174–177.
Brown C (2014) *Postcards from the County.* Pros and cons. Available at: www.prosandcons.com.
Brown G (1850) Address of the subject of prison discipline. *The Globe and Mail*, April 30.
Bull M and Back L (2003) Introduction: Into sound. In Bull M and Back L (eds) *The Auditory Culture Reader*. Oxford; New York: Berg, 1–18.
Cabot R (1978) Impulse response testing of acoustic spaces. *Acoustics, Speech, and Signal Processing, IEEE International Conference on ICASSP '78* 3: 820–823.
Cameron L (2014) Participation, archival activism and learning to learn. *Journal of Historical Geography* 46: 99–101.
Canada (1934) *Penitentiary Regulations 1933 [Penitentiary Act]*. Ottawa: King's Printer.
Canada Commission of Inquiry into Certain Disturbances at Kingston Penitentiary during April 1971, Report [headed by Justice JW Swackhamer] (1973) Ottawa: Information Canada.
Canada Department of Justice (1933) *Report of the Superintendent of Penitentiaries: Kingston Penitentiary Disturbances, 1932*. Ottawa: King's Printer.
Canada Provincial Penitentiary Commission [Brown Commission] (1849) *Report of the Commissioners Appointed to Inquire into and Report Upon the Conduct, Economy, Discipline and Management of the Provincial Penitentiary.* Montreal: Rollo Campbell.
Caron R (1978) *Go-boy!: Memoirs of a Life Behind Bars.* Toronto; New York: McGraw-Hill Ryerson.
Caron R (1985) *Bingo!* Toronto: Metheun.
Cluett S (2010) Acoustic projection and the politics of sound. *Center for Arts and Cultural Policy Study. Working Paper Series* 41: 1–28.
Cluett S (2013) Loud speaker: Towards a component theory of media sound. Unpublished doctoral dissertation, Princeton University.

Correctional Service of Canada (2010) *Corrections in Canada: An Interactive Timeline.* Available at: www.csc-scc.gc.ca/hist/1900/index-eng.shtml.

Crewe B (2011) Depth, weight, tightness: Revisiting the pains of imprisonment. *Punishment and Society* 13(5): 509–529.

Curtis D, Graham A, Kelly LA, and Patterson A (1985) *Kingston Penitentiary: The First Hundred and Fifty Years: 1835–1985.* Ottawa: Correctional Service of Canada.

Dominion of Canada (1933) *Annual Report of the Superintendent of Penitentiaries.* Ottawa: King's Printer.

Dominion of Canada (1934) *Annual Report of the Superintendent of Penitentiaries.* Ottawa: King's Printer.

Evans R (1982) *The Fabrication of Virtue: English Prison Architecture, 1750–1840.* Cambridge: Cambridge University Press.

Feld S (1996) Waterfalls of song: An acoustemology of place resounding in Bosavi, Papua New Guinea. In *Senses of Place.* Seattle: University of Washington Press, 91–135.

Foley M and Lennon J (1996) Editorial: Heart of darkness. *International Journal of Heritage Studies* 2(4): 195–197.

Foucault M (1977 [1975]) *Discipline and Punish: The Birth of a Prison.* New York: Vintage Books.

Gallagher M (2011) Sound, space and power in a primary school. *Social and Cultural Geography* 12(1): 47–61.

Gallagher M (2014) Sounding ruins: Reflections on the production of an "audio drift." *Cultural Geographies*: 1–19. Available at: doi: 10.1177/1474474014542745 (accessed 25 August 2014).

Gallagher M and Prior J (2014) Sonic geographies: Exploring phonographic methods. *Progress in Human Geography* 38(2): 267–284.

Goodman S (2010) *Sonic Warfare: Sound, Affect, and the Ecology of Fear.* Cambridge: Massachusetts Institute of Technology Press.

Guenther L (2013) *Solitary Confinement: Social Death and its Afterlives.* Minneapolis: University of Minnesota Press.

Hennessy P (1999) *Canada's Big House: The Dark History of the Kingston Penitentiary* Toronto: Dundurn.

Ignatieff M (1978) *A Just Measure of Pain: The Penitentiary in the Industrial Revolution, 1750–1850.* New York: Pantheon Books.

Jackson M (1983) *Prisoners of Isolation: Solitary Confinement in Canada.* Toronto: University of Toronto Press.

Jacobson C, Jacobson J, and Crowe T (1989) Hearing loss in prison inmates. *Ear and Hearing* 10(3): 178–183,

James G (2014) *Inside Kingston Penitentiary: 1835–2013.* London: Black Dog.

Jewkes Y (2002) *Captive Audience: Media, Masculinity and Power in Prisons.* Cullompton, Devon: Willan.

Kanngieser A (2012) A sonic geography of voice: Towards an affective politics of voice. *Progress in Human Geography* 36(3): 336–353.

LaBelle B (2010) *Acoustic Territories: Sound Culture and Everyday Life.* New York: Continuum.

Library and Archives Canada (1943) RG 73, MF T-1943. *Kingston Penitentiary Punishment Record Books.*

Maitland S (2008) *A Book of Silence.* London: Granta Publications.

Marr CA (1998) "A series of nasty situations": The causes and effects of riots at Kingston Penitentiary. Unpublished Master's thesis, Queen's University.

Martin L and Mitchelson M (2009) Geographies of detention and confinement: Interrogating spatial practices of confinement, discipline, law and state power. *Geography Compass* 3(1): 459–477.

Moran D (2012) "Doing time" in carceral space: TimeSpace and carceral geography. *Geografiska Annaler B* 94(4): 301–316.

Moran D (2013a) Carceral geography and the spatiality of prison visiting: Visitation, recidivism, and hyperincarceration. *Environment and Planning D: Society and Space* 31(1): 174–190.

Moran D (2013b) Between outside and inside? Prison visiting rooms as liminal carceral spaces. *GeoJournal* 78(2): 339–351.

Moran D, Pallot J, and Piacentini L (2011) The geography of crime and punishment in the Russian Federation. *Eurasian Geography and Economics* 51(1): 79–104.

Morin K (2013a) Distinguished Historical Geography lecture: Carceral space and the usable past. *Historical Geography* 41: 1–21.

Morin K (2013b) "Security here is not safe": Violence, punishment, and space in the contemporary US penitentiary. *Environment and Planning D: Society and Space* 31(3): 381–399.

Oliver P (1998) *"Terror to Evil-doers": Prisons and Punishments in Nineteenth-century Ontario*. Toronto: University of Toronto Press.

Ong WJ (2002) *Orality and Literacy: The Technologizing of the World*. London: Routledge.

Paglen T (2006) Recording carceral landscapes. *Leonardo Music Journal* 16: 56–57.

Parr J (2010) *Sensing Changes: Technologies, Environments and the Everyday, 1953–2003*. Vancouver: UBC Press.

Philo C (2012) Security of Geography/Geography of security. *Transactions of the Institute of British Geographers* 37: 1–7.

Piché J (2011) "Going public": Accessing data, contesting information blockades. *Canadian Journal of Law and Society* 26(3): 635–643.

Piché J (2012) Accessing the state of imprisonment in Canada: Information barriers and negotiation strategies. In Larson M and Walby K (eds) *Brokering Access: Politics, Power, and Freedom of Information in Canada*. Vancouver: UBC Press, 234–260.

Piché J and Walby K (2010) Problematizing carceral tours. *British Journal of Criminology* 50(3): 570–581.

Pizarro JM, Stenius VML, and Pratt TC (2006) Supermax prisons: Myths, realities, and the politics of punishment in American society. *Criminal Justice Policy Review* 17: 6–21.

Rogalsky M [producer] (2014) *For Those Who Stay*. PS I Love You (digital album).

Queen's University Archives (QUA) (1835–1853) MF 2280.1. Kingston Penitentiary fonds. *Punishment Record Books*. Series One, Book One: December 30, 1835 to May 3, 1853; (1843–1869) Series One, Book Two: November 11, 1843 to March 31, 1869; (1850–1856) *Female*. Series One, Book Three: December 26, 1850 to December 1856.

Queen's University Archives (QUA) (1952) V 25.5 30–262.2. George Lilley fonds. *CKWS Radio Broadcast*, May.

Ricciardelli R, Crichton H and Adams L (2014) Stuck: Conditions of Canadian confinement. In Deflem M (ed) *Punishment and Incarceration: A Global Perspective*. Available at: http://dx.doi/org/10.1108/S1521-613620140000019004.

Rice T (2003) Soundselves: An acoustemology of sound and self in the Edinburgh Royal Infirmary. *Anthropology Today* 19(4): 4–9.

Rice T (2013) Broadcasting the body: The "private" made "public" in hospital soundscapes. In Born G (ed.) *Music, Sound, and Space: Transformations of Public and Private Experience*. Cambridge: Cambridge University Press, 169–185.

Rosenzweig R and Thelen D (1998) *The Presence of the Past: Popular Uses of History in American Life*. New York: Columbia University Press.

Safe Streets and Communities Act (2011) Bill C-10, 41st Parliament of Canada, 1st session.

Sapers H (2013) *Annual Report of the Office of the Correctional Investigator of Canada 2012–2013*. Ottawa, Ontario: Office of the Correctional Investigator, Public Works and Government Services Canada.

Schafer RM (1977) *The Tuning of the World*. New York: Knopf.

Schafer RM (1994) *The Soundscape: Our Sonic Environment and the Tuning of the World*. Rochester, VT: Destiny Books.

Sennett R (1994) *Flesh and Stone: The Body and the City in Western Civilization*. New York: W.W. Norton.

Smith SJ (1997) Beyond geography's visual worlds: A cultural politics of music. *Progress in Human Geography* 21(4): 502–529.

Stansfeld S and Matheson M (2003) Noise pollution: Non-auditory effects on health. *British Bulletin* 68: 243–257.

Sterne J (2003) *The Audible Past*. Durham, NC: Duke University Press.

Sui D (2000) Visuality, aurality, and shifting metaphors of geographical thought in the late twentieth century. *Annals of the Association of American Geographers* 90(2): 322–343.

Sykes GM (1958) *The Society of Captives: A Study of a Maximum Security Prison*. Princeton, NJ: Princeton University Press.

Thomson HCH and Macaulay J (1832). Report of the Commissioners appointed ... for the purpose of obtaining Plans and Estimates of a Penitentiary to be erected in this province. *Journal of the House of Assembly of Upper Canada (1832–3)*.

Upper Canada Legislature. House of Assembly (1836) Appendix, Report of the Penitentiary Inspectors (No. 10) Rules and Regulations of the Penitentiary. *Journal of the House of Assembly (1836–7)*: 19–26.

Wener RE (2012) *The Environmental Psychology of Prisons and Jails: Creating Humane Spaces in Secure Settings*. Cambridge: Cambridge University Press.

Westerkamp H (1974) Soundwalking. *Sound Heritage* 3(4): 18–27.

Yeager MG (2008) Getting the usual treatment: Research censorship and the dangerous offender. *Contemporary Justice Review* 11(4): 413–425.

3 The prison inside

A genealogy of solitary confinement as counter-resistance

Brett Story

Introduction

In the mid-1980s political prisoner Susan Rosenberg was transferred, along with two others, into an experimental solitary confinement wing called the High Security Unit situated in the basement of the Federal Correctional Institution in Lexington, Kentucky. The extreme austerity and isolation seemed to Rosenberg even then as a harbinger of things to come:

> We always said that if they can do that to us … this will open the door to the use of these kinds of techniques and tactics for the rest of the population. And I think that has really been what has happened. It led the way to a much broader use of these kind of control management units, or solitary confinement.
>
> (Rosenberg, 2013)

Rosenberg's trepidations about the normalization of penal isolation units have indeed proven prescient. Beginning in the 1960s and proliferating most intensively over the past 30 years, we have seen the use of solitary confinement expand in the United States as an increasingly systematic, long-term practice within the nation's prisons. The practice of solitary confinement has proliferated perhaps most ferociously through the establishment and construction of control units and super-maximum (or 'supermax') prisons. Massive, highly technologized, and largely impenetrable to members of the public, supermax prisons are specifically designed for the prolonged and often indefinite isolation of prisoners classified administratively as high risk and/or problematic in some way. Control units, likewise, are the designated part of a prison that operates under a 'super-maximum security' regime, characterized by the indefinite lockdown of prisoners in conditions of solitary confinement for 22 to 24 hours a day (Shalev, 2011).

Many thousands of prisoners have endured years or even decades isolated within such edifices, and it is to the question of their relationship to the longer standing practice of isolating prisoners that I seek to contribute in this chapter. While there was one control unit prison in the United States in 1985, by 1995 there were more than 40. Today every state has its own supermax prison. Current estimates suggest there are at least 80,000 prisoners in isolated confinement on any given day in America's prisons and jails, including some 25,000 in long-term solitary in supermax prisons, with reason to believe that the numbers are actually even higher (Casella and Ridgeway, 2012). Since the early 1980s isolation has constituted one of the fastest-growing conditions of detention. While the overall prison population increased by 28 percent from 1995 to 2000, the number of US prisoners in solitary grew by 40 percent during that time – outpacing, in other words, the prison buildup itself (Johnson, 2010).

The basic question I ask here is: *why*? Specifically, why has solitary confinement expanded *now*, so systematically, for so many people and for such long periods of time? This chapter offers a new framework for making sense of the dramatic proliferation of solitary confinement during the contemporary period of US penal history. Deploying a genealogical approach, I reorient the frame of inquiry to place at the forefront isolation's effects on the *social* body in captivity, rather than the *individual* subject conventionally assumed to constitute its primary target. My objective in approaching this history genealogically is not to trace a straight and causal line between the events recounted and the mass expansion of long-term isolation, but rather to excavate a historical *field* within which a conjunction of multiple struggles and plays of force conditioned the emergence of the solitary confinement regime we witness today.

I draw upon a variety of primary sources, including interviews, media reports culled from the New York State Archives and transcripts of historical events, as well as offering a new interpretation of secondary literatures, to argue that solitary confinement operates as a socio-spatial fix to the systemic problem of prisoners' social power. The social power of prisoners manifested most spectacularly in the crisis that gripped the carceral state during the prison rebellion years of 1967 to 1972. It is out of the confluence of crises of this period, I argue, that the now generalized regime of solitary confinement – in the form of control units and supermaxes but also in the long-term isolation of many prisoners in non-supermax prisons – that the practice of pre-emptive counter-resistance was born. Critically investigating the effects of isolation on the social life and collective power aggregated by the prison, I argue, helps us recuperate the *productive* work done by solitary confinement, and allows us to both make better sense of its dramatic proliferation and potentially interrupt it.

Solitary confinement's "three waves": a counter framework of analysis

In recent years a kind of "spatial turn" has taken place in the study of the US prison system, with critics of the carceral state beginning to couch their inquiries in increasingly geographic terms. Ruth Wilson Gilmore, for example, characterizes prisons as "partial geographic solutions to political economic crisis, organized by the state, which itself is in crisis" (2007: 26). Carceral geography is an emerging field out of which we have seen new and important attention paid to the spatialities of prison siting (Bonds, 2006; Gilmore, 2007), prison visitation (Moran, 2013), state restructuring (Peck, 2003), and urban social control

(Wacquant, 2009). Such spatial thinking about prisons has not, however, been seen as having much application to the study of solitary confinement units, with some exceptions (see Morin, 2014). Yet penal isolation is, almost by definition, a spatial tactic, one that does its work on the very basis of a spatial remove of one prisoner from another. Solitary confinement is traditionally defined as a form of incarceration where prisoners spend 22 to 24 hours a day alone in a 6- by 9-foot cell in separation from each other, with no opportunity for human contact or communication.

Most accounts periodize solitary confinement's history into three "waves," tending to treat them as not only discrete and unrelated, but also attributable to the differing theories of justice assumed to animate penal policy during each period. The first wave coincided with the emergence of the penitentiary itself as a system of punishment in the early nineteenth century, during which the solitary cell was imagined as a *redeeming* space for the transformation of the prisoner into an ideal citizen-subject. Originating with humanists and Quakers in Pennsylvania, the cellular isolation of prisoners, also known as the "separate system," was championed as an enlightened corrective to pre-existing corporeal punishment practices. In the one-man, totally silent cells, it was believed, criminals would discover inside of themselves the remorse and self-realization to self-govern (Smith, 2009). In practice however, total isolation proved to do the very opposite: it drove prisoners *mad*. As early as the 1830s, reports began to document the various mental disorders prisoners were exhibiting, including hallucinations, dementia, and "monomania" (Guenther, 2013). By the late 1800s the concept of total isolation had been thoroughly discredited. Solitary confinement was redesigned as a short-term punishment for misbehaving prisoners, and retreated for almost a century from common use.

Solitary confinement was revived as a systematic practice in US prisons beginning in the 1950s. This "second wave" spanned the 1950s to the 1970s, during which a programmatic deployment of isolation in prisons was introduced as a strategy of behavior modification. Its piecemeal resurgence during this period was led by behavioral scientists and prison officials keen to apply tactics developed internationally, in the wake of the Korean War. They targeted particularly problematic domestic prisoners in the US to treat and recondition their "undesirable" behavior (see Gómez, 2006; Smith, 2009; Guenther, 2013).

The "third wave" of the practice has been marked by the massive proliferation and long-term use of such spaces via designated "control units" within existing institutions, and the building of new structures designed to hold all their inhabitants in indefinite isolation. This period began in the 1980s and saw solitary confinement divested of either its redemptive or its rehabilitative ideals. During this period the number of prisoners kept in long-term isolation exploded, most dramatically under the aegis of "supermax" prisons, which – despite being extremely expensive to both construct and manage – proliferated across the United States throughout the 1990s and 2000s (Shalev, 2011: 153). It is this "third wave" in the history and historical geography of solitary confinement that scholars have found most difficult to explain. Despite the massive costs and deleterious mental health effects of long-term penal isolation, the literature investigating the purpose of control units and their structural functions within the contemporary prison system is still sparse (Mears and Reisig, 2006: 34). Few researchers have attempted to look systematically and historically at the specific justifications, uses, or broader social consequences of institutionalized solitary confinement, let alone also the "structured anxieties, political contradictions, and constitutive crises" (Rodriguez, 2006: 43) through which control and supermax units have cohered socio-historically.

Most scholars tend to explain current practices in one of two ways. The first argument is that institutionalized, long-term isolation constitutes a purely punitive tool associated with a general hardening of the prison system writ large; a further excess of a system itself putatively defined by a "culture" of brutality, or a "punitive mentality" (Kamel and Kerness, 2003; Haney, 2008). This argument tends to presume that the practice is structured less by a productive logic or political economy, and more by prison *culture* or the disposition of prison staff. Such arguments, in general, thus fail to shed explanatory light on the actual construction, funding, and design of such units at this particular historical conjuncture, tending to invoke a kind of affective pathology at the expense of structural analysis.

The second explanation characterizes the expansion of control units as a response to increasing violence inside of prisons; the purpose thus being to "stem violent behavior" or to control "the worst of the worst" prisoners (King, 1999; Kurki and Morris, 2001; Ward and Werlich, 2003). Most of this work, however, tends to uncritically reproduce corrections officials' own claims and rationalizations about increased levels of disorder in the prison system, rather than offering critical investigation into isolation's violence-curbing effects. Indeed, what evidence does exist suggests that control units do little, if anything, to generate order or stem violence (Mears and Reisig, 2006). One study concluded that "the effectiveness of supermax prisons as a mechanism to enhance prison safety remains largely speculative" (Briggs *et al.*, 2003: 1371).

This chapter offers another interpretation of the history of solitary confinement in the US: one that reframes the very object of the isolation unit's spatial attention and asks not what this spatial practice *fails* to do but what it *succeeds* in doing. I suggest that attempts to make sense of the surge of prison isolation have been misled by the presumption of the *individual* as the object of isolation. While that presumption seems self-evident, given that it is the confinement of the individual body that isolation's very architecture, this presumption serves only to make the practice *il*legible, given its well-evidenced deleterious effects on prisoners' impulse control, among other things (for a discussion on how such units produce *more* prison violence, for example, see Morin, 2013). The apparent illogics and failures of mass solitary confinement thus suggest the need for an alternative framework: that we consider this antisocial penal practice as, in fact, targeting a *social* body held captive by the prison edifice. I suggest that analyzing the solitary confinement unit through the frame of a *social* subject rather than *individual* subject helps reveal and recuperate its productive work and systemic logic within the prison system writ large.

In the following section I trace the historical origins of the modern control unit within the broader context of prisoner resistance to the US carceral regime through the latter half of the twentieth century. Rather than historicizing solitary confinement in three discrete carceral eras and then trying to make sense of them as if accountable to divergent logics, I adopt a genealogical approach, mapping out the continuities as well as the ruptures between these three periods. My use of genealogy here borrows from prison scholar Dylan Rodriguez (2006: 43), who contends that a:

> critical or radical genealogy of the prison regime, as opposed to a traditional history or sociology of the prison as a discrete institution … foregrounds the processes and struggles through which the coherence of this categorical arrangement – the grounding of the prison's domination – is restored through and from varieties of historical crisis.

I heed his call to pay fundamental theoretical attention to the production *and* productivity of the regime of penal isolation, through the constitutive crises and structured anxieties that surround and characterize its socio-historical ascendance. I argue that it is in response to, and pre-emption of, the *social crisis* posed by aggregated and racialized prisoner bodies, and their collectivization at the site of captivity, that the solitary confinement cell functions as a socio-spatial "fix" against internal threats to the carceral state.

USP-Marion, Black Muslims, and the behavior modification experiments

One place to begin the story of the modern control unit is at the US Penitentiary at Marion, Illinois on July 17, 1972. On that day, a multiracial coalition of radical prisoners organized a prison-wide work stoppage in response to the severe beating of a Chicano prisoner named Jesse Lopez by a prison guard. Two days after the attack they issued a call to their fellow prisoners to collectivize themselves into an organized political action:

> We: the concerned prisoners ask every prisoner to cooperate in a general work stoppage, if Jesse Lopez is not released by 12:00 today (7–17–72) and if assurance of the prosecution of the officer in question is not forthcoming.
> (quoted in Gómez, 2006: 72–73)

Almost 300 prisoners heeded the call and went on strike. In response, prison officials first locked down *all* prisoners in their cells for six days, and then continued to hold the seven suspected strike leaders in indefinite isolation. The strike was suspended briefly but resumed within weeks. Officials again responded by segregating prisoners, this time locking *60* men into solitary confinement and enrolling them in the prison's "behavior modification" program, known as CARE (Control and Rehabilitation Effort). Located in the prison's H-Unit, the

CARE program's main tactic of behavior modification was the use of solitary confinement for an indefinite period of time. In 1973 the H-Unit at Marion was officially designated the Long-Term Control Unit, the first documented use of that term in US penal history (*Adams v. Carlson*, 1973: 621–622).

United States Penitentiary (USP) Marion had constituted, from its very beginning, a politically significant prison. Opened in 1963, the prison was constructed to hold 500 "adult male felons who are difficult to control" (Mitford, 1974: 199), quickly becoming a warehousing space for some of the most politically active, racially conscious, and organized prisoners from around the country (Gómez, 2006: 59). Their activities, often coordinated through cross-racial coalitions, included organizing clandestine ethnic studies classes, striking against unsafe prison factory conditions, teaching each other to become jailhouse lawyers, and collaborating with allies outside to draw public and international attention to brutal prison conditions. Following the guards' attack on Jesse Lopez, these variously affiliated prisoners began to organize under the banner of the Political Prisoners Liberation Front (PPLF), and found strength in their common alliance as poor and racialized people in captivity. As one prisoner put it, "The more we developed and joined hands across color lines, the more we became a threat" (quoted in Gómez, 2006: 58). As Alan Eladio Gómez writes (2006: 59),

> Identifying these prisoners as leaders or "problem inmates", prison authorities contended that by isolating them in the same institution and employing a series of behavior-modification techniques, as well as physical and psychological torture, they could control dissent.

Although it opened in 1963, most transfers to USP-Marion did not actually take place until the late 1960s (Breed and Ward, 1985: 10), which is also when the prison began experimenting systematically with solitary confinement. The CARE program, which was first implemented in 1968, relied on solitary confinement as its primary disciplinary tactic. The program was developed by Marion's prison psychologist Martin Groder in collaboration with a seminal researcher on Chinese and Communist brainwashing tactics, named Dr. Edgar Schein. Schein was a central figure in the adaptation of total isolation as a tactic of behavior modification for application inside prisons during the 1960s. His influence and ideas may be traced in the published proceedings of a pivotal symposium organized by the U.S. Bureau of Prisons in 1961 called "The Power to Change Behavior," for which Schein was the invited keynote speaker. The seminar was organized to bring together prison wardens and behavioral scientists working at the cutting edge of research on behavior reform and brainwashing techniques developed in prisoner-of-war camps during the Korean War. Transcripts of its proceedings illuminate institutional vulnerabilities for which solutions were being sought.

Dr. Schein was invited to present his theories on brainwashing and their applicability to the context of US prisons. He told his audience:

> I would like to have you think of brainwashing not in terms of politics, ethics, and morals, but in terms of the deliberate changing of behavior and attitudes by a group of men who have relatively complete control over the environment in which the captive population lives.
>
> (U.S. Bureau of Prisons, 1961: 56)

The most important implication of his findings for a general theory of "attitude change," Schein argued, is that dismantling social support and opportunities for social contact is crucial. To that end he suggested a set of key techniques for prison wardens to try on their problem prisoners: (1) to physically remove the prisoner to an area sufficiently isolated in order to break or seriously weaken close emotional ties; (2) to segregate all natural leaders; (3) to prohibit group activities that do not fit brainwashing objectives; (4) to systematically withhold mail; (5) to create a feeling among the isolated group of prisoners that they have been abandoned by and totally isolated from the community; (6) to undermine all emotional supports; and (7) to preclude access to literature which does not aid in the brainwashing process (Elijah, n.d.).

Schein's remarks on the use of isolation and segregation for behavior and "attitude change" among prisoners become particularly significant in the broader context of the symposium's proceedings. A persistent subject during the symposium was the "problem" of Black Muslims within the nation's prisons during this period. In his written summary of the conference proceedings, Bertram S. Brown writes: "There were specific, practical management issues raised, such as 'How shall we manage the Muslims?' 'Who should we isolate?' etc." (U.S. Bureau of Prisons, 1961: 65). He later describes Muslims as one of the main groups of concern: they represented "[p]roblems not only of difficult and recalcitrant individuals but difficult *groups* in the prison setting such as the 'Black Muslims' and in the past 'conscientious objectors'" (U.S. Bureau of Prisons, 1961: 68). Characterizing it as a discussion of "technical management," Brown referenced a particular exchange between a few of the panelists and audience members, including James V. Bennett, then Director of the BOP (Bureau of Prisons):

> I think your question, Mr. Bennett, was not so much whether you take action against the Muslims as a group, but how can you counteract the effects of the kinds of techniques they use to recruit members, etc. and cause general mischief in the prison system.
>
> Mr. Bennett: Well, the prisons have been filled with such groups at different times. During the war we struggled with the conscientious objectors – non-violent coercionists – and believe me, that was really a problem. Every day they got together as a group and put sand in the grease boxes and refused to eat and went on hunger strikes and agitated, etc., and we were always trying to find some way in which we could change or manipulate their environment.
>
> (U.S. Bureau of Prisons, 1961: 68)

Bennett had been very impressed by Schein's recommendations, and closed the symposium by encouraging prison wardens to apply the tactics discussed at the symposium in their own institutions:

> We can perhaps undertake some of the techniques Dr. Schein discussed....
> Do things on your own – undertake a little experiment with what you can do
> with the Muslims ... do it as individuals, do it as groups, and let us know
> the results.
>
> (U.S. Bureau of Prisons, 1961: 72)

Encouraged by Bennett, prison wardens across the country widely applied behavior modification programs of spatial isolation to their institutions, under various titles. At the El Reno Reformatory in Oklahoma it was called a "Sensitivity Training Program"; at the McNeil Island Federal Prison off the coast of Washington State it was called a "Therapeutic Community"; and at the Federal Medical Center for Prisoners in Springfield, Missouri, Groder and Schein collaborated again to design the Special Treatment and Rehabilitation Training (START) program.

The specific attention paid to Black Muslims during the symposium is revealing. By 1961 Black Muslims had established themselves as a formidable – and effective – political force inside America's prisons. Many of them incarcerated for their opposition to the Vietnam draft, their earliest collective organizing efforts targeted racial segregation inside prisons, building on the Civil Rights Movement more broadly in its desegregation campaigns. Between 1961 and 1978 there were 66 reported federal decisions involving Black Muslims (Gottschalk, 2006: 175). The judicial decisions that came out of these lawsuits gave prisoners unprecedented protections and rights behind bars, including the right to provide legal aid to one another. Perhaps even more significantly, their demonstration of an ethos of group solidarity, one based on the notion that Blacks as a group faced collective oppression, and their ability to execute a disciplined, effective strategy, provided a model for *collective* organizing inside prisons and laid the groundwork for subsequent prisoner struggles (Gottschalk, 2006; Berger, 2010).

It is particularly important to note that the development of penal isolation re-emerged during this period as a technique of behavior modification targeting this collectively organized group; it alerts us to the anti*social* function of solitary confinement inside prisons. The establishment of such behavior modification programs thus prefigured the expansion of penal isolation as a deliberate technical and architectural tactic of *disassociation*, focused primarily on prisoners whose collective activities demonstrated the possibilities of solidarity, mutual aid, and social power among oppressed people. Even after the decline of the Nation of Islam behind bars in the mid-1960s, the presence of a growing and disproportionate number of Blacks, Puerto Ricans, and other racialized groups, at the same time as rising political mobilization and tension around racial issues outside the prison system, helped focus attention on prisons in a new and powerful way. This period would become known as the "prison rebellion years," and the disproportionate use of isolation against racialized and politicized prisoners during and after this period is the context from which today's super-max emerged.

Attica and the "prison rebellion years"

The years between 1967 and 1972 saw an explosion in confrontations staged by prisoners against the state (see Cummins, 1994; Berger, 2010; Mirpuri 2010). Gottschalk (2006) counts 132 riots in US prisons during this period, subsequently dubbed the "prison rebellion years" (Gómez, 2006: 58). This period also saw an explosion in the number of organizations dedicated to reforming or eliminating prisons, many of which used the space of the prison itself as an organizing base. As well as litigating and organizing work stoppages, prisoners set up underground reading groups, elaborate education systems, prison newspapers, and even labour unions. By the 1970s, Dan Berger argues, a new kind of political movement had developed in the US that specifically "targeted the prison as an institution ... and used the prison system as the launching point for a critique of the institutions of America's racial order" (2010: 32).

While a review of this rich and varied history exceeds the boundaries of this chapter, there are two points that are important to make about the prison rebellion years, and for which I will offer a few salient examples culled from the archives. The first is that the collective organizing inside and across prison walls during this period posed a deeply registered crisis for state officials and the prison regime. Second, while supermaxes and similar penal prototypes for mass isolation did not proliferate as carceral forms until more than 15 years later, it is in the aftermath of this period's prisoner-led collective organizing that one sees the spatial tactic of mass penal isolation first become flagged publicly, and then begin to gestate politically.

On September 9, 1971, prisoners at Attica Correctional Facility in upstate New York erupted in what remains the largest prison rebellion in US history. In a show of unprecedented cross-racial organizing, more than 1,200 of Attica's 2,432 prisoners took over one of the prison's four yards and held 42 corrections officers and civilian employees hostage. They held the prison in relative calm for 97 hours, after which Governor Nelson Rockefeller ordered the National Guard to crush the rebellion. 41 men were killed, including 31 prisoners, all of them by bullet wounds discharged by the national guardsmen and state troopers. No prisoners were found to have had firearms.

While the history of the Attica rebellion is well known, less has been written about how seriously state officials registered the uprising as a crisis for the system as a whole, and began to consider possible long-term solutions. Officials scrambling to recuperate order and control in the aftermath of Attica began to float one idea in particular: a separate, high-tech and super-maximum security prison designed to isolate politicized prisoners and prison leaders from other prisoners and from each other. Governor Rockefeller was quick to ascribe the

Attica revolt to the "revolutionary tactics of militants" (Farrell, 1971a: 1, 30), suggesting on record that the prisoners' demands "had political implications beyond the reform of the prison" (Farrell, 1971b: 1, 48). According to prison officials and their sympathizers, the risk posed by prison militants could be remedied in only one way: penal isolation. In late September State Commissioner of Corrections Russell G. Oswald used a press conference to peddle the idea of a separate, high-security facility for "problem" prisoners. Oswald told reporters that his department was seeking a facility for as many as 500 of the most troublesome prisoners, to keep them separated from what was at that time a population of 16,000 prisoners in the state's prison system. He credited Vincent R. Mancusi, superintendent of the Attica prison, who had first raised the idea of isolating five prisoners who earlier in July had signed a manifesto demanding changes in prison procedure. "Mr. Mancusi thought they were real troublemakers," Mr. Oswald recalled. "He felt they were a behavior problem and that by transferring them, the rest of the institution might not get infected" (quoted in Kaufman, 1971: 1).

As the opening witness in a five-day committee inquiry into prison disruptions later that fall, Mancusi publicly called for the establishment of a separate institution, sited in a geographically remote area, which would house "the hardcore revolutionaries" (Ringle, 1971). Members of the media and religious groups also contributed to the growing chorus on the necessity for a separate and remote "super"-facility. In an editorial in *The Daily News* from September 21, 1971, the editors laid the blame squarely on the shoulders of racialized revolutionaries. "It appears that there are four categories – the Black Panthers, the Black Muslims, the Young Lords, and those who are plain ordinary prisoners putting in their time," whose quest for power, they argued, laid the groundwork for the September 9 uprising. The editorial's proposed solution was by now a familiar one: "The impetus now must be on the future and a more realistic way of coping with this unrest.... Separate facilities for the hard core radicals are a must" (Power Struggle, 1971). At a memorial for guards killed at Attica, the prison's Roman Catholic Chaplain, Fr. James Collins, also called for the construction of a separate "maximum security institution for about 150 hardcore, militant, Marxist revolutionaries." If these individuals were not isolated, he predicted, the result would be more rebellions just like Attica's (McCandlish, 1971: 31).

Such calls continued into the fall and winter. On December 10, 1971, Oswald gave an hour-long address to some 200 members of the group Women in Support of State Correctional Employees, many of them wives of prison guards. During the speech he continued to shore up the category of the "hardcore revolutionary," as well as to suggest, tellingly, that the condition of captivity *lent itself* to the development of a political consciousness and therefore participation in rebellion.

"Correctional institutions nowhere are mansions of merry men," said Oswald. "I don't know where would be a better place for hardcore revolutionaries to recruit than in a prison set up. There seems to be some evidence that this is going on." Oswald then marshaled this warning into the case for the segregation of politically conscious prisoners, using the language, somewhat ironically, of prisoners' "rights." It was the "most precious right," he said, of every prisoner "to be protected from those in the prison population who would recruit him – who would use him – would insist that he follow their line of reasoning" (Oswald Stresses Gains, 1971).

The Attica uprising constituted a watershed moment for penal policy in the United States. New York State's own later entrées into mass isolation and the supermax model were born in its wreckage, even while the immediate aftermath saw a protracted ideological struggle between a rehabilitative orientation to penal policy and a more punitive one (Morrell, 2012: 105). While in the short-term improvements would be made to food services and other basic prison conditions, the long-term legacy of the Attica rebellion would prove to be the expansion of a solitary confinement infrastructure. New York's first "maxi-maxi prison," Southport Correctional Facility, first took root when proposed by the 1983 New York State Republican Task Force on Correctional Crisis, where it was imagined as a facility to hold prisoners deemed "troublesome" at a single location under austere and isolating conditions.

Social crisis had engendered the ultimate of antisocial technology in response. Before Southport could be built however, USP-Marion would prove once again to be a laboratory for new carceral designs.

USP-Marion and the birth of the supermax

In 1979 USP-Marion was designated a level six prison – the first and only such classified institution in the country. A Bureau of Prisons document published that same year also flagged the idea of turning the whole of USP-Marion into a "closed-unit operation" – in which all prisoners would be held in indefinite isolation – an idea further solidified into written plans described in a 1981 report (Breed and Ward, 1985: 11, 22). As punitive conditions intensified in response to political agitation in the prison throughout the late 1970s and early 1980s, so did the resistance of prisoners. A particularly intractable work strike was initiated in September 1980 when the acting warden refused to respond to a list of prisoners' demands, including the right for Native Americans to practice purification rites and Muslims to wear the fez and turban. Reputed to be the longest and most peaceful strike in prison history, the strike was never officially broken, as the administration decided instead to close the prison factory down entirely in January 1981 (Breed and Ward, 1985: 12; Lassiter, 1990: 76). Political activity continued throughout that spring and summer.

In October 1983 incarcerated members of the Aryan Brotherhood killed two correctional officers and one prisoner. Although the three men responsible for the guards' deaths were identified almost immediately, *all* prisoners were locked into their cells permanently and indefinitely following this incident. This prolonged "emergency" lockdown would come to constitute a large-scale experiment in solitary confinement that lasted at Marion for 23 years.

This particular incident became the pretext for turning Marion into the "closed-unit operation" considered by the BOP at least as far back as 1979, and ending only when the federal government converted USP-Marion into a medium-security prison in 2007 (Kim *et al.*, 2012: 11). A former guard at Marion, David Hale, testified in 1984 in front of a US magistrate in a class-action suit brought by prisoners that Marion's warden at the time actively sought to *promote* tensions and unrest, in order to produce a justification for permanently locking down the entire population in isolation (Lassiter, 1990: 27).

In a process commonly dubbed 'Marionization' (King, 1999; Ward and Werlich, 2003), the mass cellular segregation of prisoners at USP-Marion became a model for supermax prisons and control units across the country: at McAlester, Oklahoma in 1985; at Pelican Bay, California and Baltimore, Maryland in 1989; and at Southport, New York in 1991 (Raab, 1991). Southport Correctional Facility in the town of Elmira, in upstate New York, is the fruition of the separate, maximum-maximum security prison first championed by prison officials in the wake of the Attica rebellion. Built as a maximum-security facility in 1988, Southport underwent a further conversion in early 1991 to what the state calls a "punitive segregation" facility, much like USP-Marion, characterized by the holding of all prisoners in isolation cells called Special Housing Units. All 600 prisoners at Southport today are held in isolation for 23 hours a day, with an hour for recreation in a metal exercise pen (Morrell, 2012: 202).

The specter of crisis and solitary confinement as antisocial fix

Gilmore argues that "crisis means instability that can be fixed only through radical measures, which include developing new relationships and new or renovated institutions out of what already exists" (2007: 26). While the late 1960s saw the emergence of a particular historical conjuncture that rendered the prison and its captives as key focal points for Black liberation and other freedom struggles, it is possible to argue that the specter of political uprising haunts the very condition of captivity itself. Dylan Rodriguez is among those who argue that the crisis which prisoner struggles pose is endemic, rather than circumstantial, to the prison's regime of "human immobilization and bodily disintegration" (2006: 7). At the same time as "the structuring of unfreedom extinguishes the possibility of legitimate political subjectivity *a priori*," it also interpellates and even "reconstitutes" the prisoner in such a way as to invest in her the basis of its own critique, opening a "pathway of radicalism and insurgency" (Rodriguez, 2006: 37, 7). In other words, the space of the prison *produces* the conditions of political subjectivity out of the same technology by which it *extinguishes* the prisoner's juridical and bodily personhood.

That the prison's system of social relations gives rise to the agents of its own potential dismantling is a bold claim. It is significant, however, that this view emerges from those examining the testimonials of prisoners themselves. Barbara Harlow takes this position when she writes:

> Penal institutions, despite, if not because of, their function as part of the state's coercive apparatus of physical detention and ideological containment, provide the critical space within which, indeed from out of which, alternative social and political practices of counterhegemonic resistance movements are schooled.
>
> (Harlow, 1992: 10)

The *sociality* of the prison context is critical: it is the recognition which prisoners experience once inside of sharing a common condition of oppression, and a *collective* aggregation of bodies and energies against the sophisticated weaponry of the state, that renders the prison such a critical site of social struggle and collectivized power.

My suggestion is that we consider the solitary confinement cell in similar terms as the prison writ large: as a partial, spatial resolution to multiple, intersecting crises and the prison's own central role in reproducing the neoliberal state. It is an antisocial fix for a social crisis; a crisis prisoners have posed, and may pose, by virtue of their collective oppression and spatially aggregated bodies. Such a framework helps us to see the connection between the use of solitary confinement from the 1950s to the 1970s against political leaders, groups, and participants *in particular*, and the wholesale structuring of segregation cells and control units into a systemic carceral infrastructure *in general* in the decades since. The connecting thread also emerges in interviews with ex-prisoners who have spent long periods in isolation, who describe how overt political activity is often pre-empted by the very *possibility* of isolation. As one ex-prisoner "David" (2013) put it:

> It's a known thing that you're not going to be an *advocate* in the prison system without going to the box. So that's squashed almost immediately.... Any person that has the mentality to try and fight against the conditions inside the prison is automatically going to be deemed a troublemaker. That person is going to be targeted.

A rereading of isolation's most recent history helps recast the familiar narrative of its origins, illuminating the ways in which, even in its original nineteenth-century "reformist" guise, the isolation cell served a deliberate function of internal social control. In texts from the period, reformers registered a deep and abiding fear that a revolutionary conspiracy might at any moment break out among the condemned. Smith (2009), for example, describes the reformers' obsession with political as well as epidemiological *contagion* within spaces of penal captivity; the same condition of bodily proximity that facilitated the spread of jail fever would also, it was believed, spread the ideology and tactics of riot and rebellion. Beaumont and Tocqueville observed it as "the contagion of mutual communications," while Jeremy Bentham lamented the "thronging," "jostling," "confederatings," and "plottings" that plagued British gaols (quoted in Smith, 2009: 87, 88). "The grave problem for reformers of the late eighteenth century," writes Smith, "was the "loathsome communion" of prisoners' bodies and souls, the conspiratorial mingling that threatened to spread from the jail to the public at large, and there to inspire open rebellion" (Smith, 2009: 88).

Conclusion

Many threads have been left out of the genealogy offered here, and this chapter by no means attempts to exhaust the terrain of what solitary confinement cells *do*. Rhodes (2004), for example, calls our attention, in a way that I do not, to the assumptions about self-control and individual choice that discourses surrounding supermax confinement help reproduce. The point, however, is that whatever else the solitary confinement cell does, it also enables the prison system, as a social formation, to both respond to and anticipate political challenges from those subjected to captivity. In expanding, institutionalizing, and normalizing the spatial architecture of isolation with the nation's prisons, state actors have also systematized a new ideological basis and set of categories (the "incorrigible prisoner," the "prison gang member," the "worst of the worst") upon which extreme isolation is legitimated *and* the prisoner's imminent or inchoate threat is pre-empted.

A genealogy of the solitary confinement cell tells us that there is indeed a logic that weaves through penal isolation's so-called three waves. To follow that logic, we must abandon the seemingly self-evident notion that the object of the solitary cell's coercive attention is the individual body around which it wraps its immediate architecture. To follow it further we must disabuse ourselves of the intellectual tendency to observe penal trends through the idealism of particular theories of justice, as if the ideas came first and the material practice followed. The connecting tissue between the three "waves" of solitary confinement in US prison history and historical geography is the administrative imperative of institutional self-preservation: the reproduction of prisons and the carceral state that operates them despite and against the resistance captivity inculcates in its subjects. Crisis in that reproduction opens up the terrain to struggle, and the refinement and expansion of the solitary confinement regime is one outcome of such struggle. But it is not the only possible outcome.

Solitary confinement and its generalization within the contemporary carceral landscape may thus be traced to the social power endemic to carceral space and the counter-resistance which such social power renders possible. The mass proliferation of isolation cells over the past few decades thus has much to tell us about the vulnerabilities of the prison regime itself. The "usable past" of the solitary confinement unit is a living history of struggle, of political agency, and of social being. Restaging this history to highlight prisoners' active resistance to the oppressive structures of the prison regime thus constitutes an important means not only of recuperating prisoners' political power, but also of reminding us that even the most extreme technologies of state violence can be challenged.

As a spatial resolution of crisis and a pre-emption of further crisis within the prison system, the "fix" of solitary confinement is a partial and certainly mutable one. The mass isolation of prisoners has opened up a new terrain of collective action, and the current conjuncture's most powerful and sustained contestations of the prison regime come from *within* and organize *across* the solitary confinement cell. California's third collectively organized prison hunger strike in two years, originating from the deepest bowels of the Pelican Bay isolation units in July 2013, came to an end that September. At its height there were at least 30,000 prisoners participating, the majority of them held in isolation for many years at a time. Expressing the seriousness with which the correctional regime has registered these strikes, a subsequent pro-prison industry event held in California in December 2013 was organized under the banner "When Hunger Shook the Penal System" (2013).

Since then there have been at least two other major hunger strikes waged by prisoners across the country protesting, most immediately, their isolated confinement (see e.g., Lynd and Lynd, 2014). Theirs will not be an easy fight, as prisoners, further fragmented and immobilized by the austerity of their isolation chambers, are left with nothing more than their own embodiment to put on the line. Their immediate carceral environment is, as Rodriguez puts it, "a sterilized, whitewashed, and state-proctored condition that magnifies the coerced rupture of the human from the social" (2006: 190). The consequences of this rupture can be and has been devastating, but it can also engender new means of forging social bonds and collective strength. As political prisoner Angela Davis once put it:

> Those of us with a history of active struggle against political repression understood of course, that while one of the protagonists in this battle was indeed the state, the other was not a single individual, but rather the collective power of the thousands and thousands of people opposed to racism and political repression.
>
> (Davis, 1971: xiii–xiv)

Prison rebellions have proven dangerous to the social formation precisely because they implicate and make legible structures of unfreedom *outside* the prison as well as inside, while forecasting the possibility of liberation struggle, and even success.

References

Adams v. *Carlson* (1973) 488 F.2d 619 (7th Cir 1973).

Berger D (2010) *We Are the Revolutionaries: Visibility, Protest, and Racial Formation in 1970s Prison Radicalism.* Unpublished doctoral dissertation, University of Pennsylvania.

Bonds A (2006) Profit from punishment? The politics of prisons, poverty and neoliberal restructuring in the rural American northwest. *Antipode* 38(1): 174–177.

Breed AF and Ward D (1985) *The United States Penitentiary Marion, Illinois, Consultants' Report Submitted to the Committee of the Judiciary, US. House of Representatives,* December 1984 in Marion Penitentiary.

Briggs CS, Sundt JL, and Castellano TC (2003) The effect of supermaximum security prisons on aggregate levels of institutional violence. *Criminology* 41: 1341–1376.

Casella J and Ridgeway J (2010) No evidence of national reduction in solitary confinement. *Solitary Watch*, June 15. Available at: http://solitarywatch.com/2010/06/15/no-evidence-of-national-reduction-in-solitary-confinement/.

Casella J and Ridgeway J (2012) How many prisoners are in solitary confinement in the United States? *Solitary Watch*, February 1. Available at: http://solitarywatch.com/2012/02/01/how-many-prisoners-are-in-solitary-confinement-in-the-united-states/.

Cummins E (1994) *The Rise and Fall of California's Radical Prison Movement.* Stanford, CA: Stanford University Press.

David (2013) Interview with author, March 13.

Davis A and other political prisoners (1971) *If They Come in the Morning, Voices of Resistance.* New York: The Third Press.

Elijah JS (n.d.) New African/Black political prisoners and prisoners of war: Conditions of confinement. *Research Committee on International Law and Black Freedom Fighters*: Freedom Archives. Available at: http://freedomarchives.org/Documents/Finder/DOC510_scans/New_Afrikan_Prisoners/510.new.african.pp.pow.conditions.confinement.pdf.

Farrell WE (1971a) Rockefeller sees a plot at prison. *New York Times*, September 14: 1, 30.

Farrell WE (1971b) Governor defends order to quell Attica Uprising. *New York Times*, September 16: 1, 48.

Gilmore RW (2007) *Golden Gulag: Prisons, Surplus, Crisis, and Opposition in Globalizing California.* Berkeley: University of California Press.

Gómez AE (2006) Resisting living death at Marion Penitentiary, 1972. *Radical History Review* 96: 58–86.

Gottschalk M (2006) *The Prison and the Gallows: The Politics of Mass Incarceration in America.* Cambridge: Cambridge University Press.

Guenther L (2013) *Solitary Confinement: Social Death and its Afterlives.* Minneapolis: University of Minnesota Press.

Haney C (2008) A culture of harm: Taming the dynamics of cruelty in supermax prisons. *Criminal Justice and Behavior* 35: 956–984.

Harlow B (1992) *Barred: Women, Writing, and Political Detention.* Hanover, NH: University of New Hampshire Press.

Johnson K (2010) States start reducing solitary confinement to help budgets. *USA Today*, June 14. Available at: www.usatoday.com/news/nation/2010-06-13-solitary-confinement-being-cut_N.htm.

Kamel R and Kerness B (2003) *The Prison Inside the Prison: Control Units, Supermax Prisons, and Devices of Torture: A Justice Visions Briefing Paper.* Philadelphia, PA : American Friends Service Committee.

Kaufman MT (1971) Oswald seeking facility to house hostile convicts. *New York Times*, September 29: 1.

Kim S, Pendergrass T, and Zelon H (2012) *Boxed In: The True Cost of Extreme Isolation in New York's Prisons.* New York: New York Civil Liberties Union.

King R (1999) The rise and rise of supermax: An American solution in search of a problem? *Punishment and Society* 1: 163–186.

Kurki L and Morris N (2001) The purposes, practices and problems of supermax prisons. In Tonry M (ed) *Crime and Justice: A Review of Research.* Chicago, IL: University of Chicago Press.

Lassiter C (1990) Robo-prison. *Mother Jones*, September/October.

Lynd A and Lynd S (2014) Illinois prisoners in Menard High Security Unit plan to begin hunger strike Jan. 15. *San Francisco BayView*, January 14. Available at: http://sfbayview.com/2014/illinois-prisoners-in-menard-high-security-unit-plan-to-begin-hunger-strike-jan-15/.

McCandlish P (1971) Prison chaplain, at guard's funeral, asks for separate facility for revolutionaries. *New York Times*, September 17: 31.

Mears DP and Reisig MD (2006) The theory and practice of supermax prisons. *Punishment and Society* 8(33): 33–57.

Mirpuri A (2010) *Slated for Destruction: Race, Black Radicalism, and the Meaning of Captivity in the Postwar Exceptional State*. Unpublished doctoral dissertation, University of Washington.

Mitford J (1974) *Kind and Usual Punishment*. New York: Random House.

Moran D (2013) Carceral geography and the spatialities of prison visiting: Visitation, recidivism, and hyperincarceration. *Environment and Planning D: Society and Space* 31: 174–190.

Morin KM (2013) "Security here is not safe": Violence, punishment, and space in the contemporary US penitentiary. *Environment and Planning D: Society and Space* 31: 381–399.

Morin KM (2015) The late-modern American jail: Epistemologies of space and violence. *The Geographical Journal*. doi:10.1111/geoj.12121

Morrell A (2012) *The Prison Fix: Race, Work, and Economic Development in Elmira, New York*. Unpublished doctoral dissertation, City University of New York.

Oswald Stresses Gains, Answers Prison Critics (1971) *Regional News Service*, December 11.

Peck J (2003) Geography and public policy: Mapping the penal state. *Progress in Human Geography* 27(2): 222–232.

Power Struggle (1971) *The Batavia Daily News*, September 21.

Raab S (1991) The region; Uprising challenges "maxi-maxi" prison idea. *New York Times*, June 2. Available at: www.nytimes.com/1991/06/02/weekinreview/the-region-uprising-challenges-maxi-maxi-prison-idea.html.

Rhodes LA (2004) *Total Confinement: Madness and Reason it the Maximum Security Prison*. Berkeley: University of California Press.

Ringle W (1971) Prisoners hatched plots by letter, Mancusi says. *Rochester Democrat and Chronicle*, November 30.

Rodriguez D (2006) *Forced Passages: Imprisoned Radical Intellectuals and the U.S. Prison Regime*. Minneapolis: University of Minnesota Press.

Rosenberg S (2013) Interview with author, June 15.

Shalev S (2011) Solitary confinement and supermax prisons: A human rights and ethical analysis. *Journal of Forensic Psychology Practice* 11: 151–183.

Smith C (2009) *The Prison and the American Imagination*. New Haven, CT: Yale University Press.

U.S. Bureau of Prisons (ed) (1961) *The Power to Change Behavior*. Papers presented at a seminar conducted by the Bureau of Prisons at an Associate Warden Training Program in April 1961. Washington, DC: Bureau of Prisons.

Wacquant L (2009) *Punishing the Poor: The Neoliberal Governance of Social Insecurity*. Chapel Hill, NC: Duke University Press.

Ward DA and Werlich TG (2003) Alcatraz and Marion: Evaluating super-maximum custody. *Punishment and Society* 5(1): 53–75.

"When Hunger Shook the Penal System" (2013) NextGen Prisons and Correctional Facilities, San Francisco. Available at: www.PrisonsCorrectionalFacilities.com.

4 "Sores in the city"

A genealogy of the Almighty Black P. Stone Rangers*

Rashad Shabazz

There they are.
Thirty at the corner.
Black, raw, ready.
Sores in the city
That do not want to heal.
(Gwendolyn Brooks, from the poem "The Blackstone Rangers," (1987))

Introduction

Black gangs in Chicago were the tempest caused by the carceral current that swept through the city's Black neighborhoods during the First World War. Carceral power was the spatial backdrop against which Black gangs like the Almighty Black Stone Rangers emerged. Architectures of confinement, restrictions on Black movement, ubiquitous policing, few job options, and rising rates of Black male incarceration provided the fertile carceral landscape for the germination of Black gangs. Some Black gangs even got their start in carceral institutions (Dawley, 1992: 10). But it was not only that carceral power was the context that gave rise to gangs; it also played a formable role in shaping their performance of masculinity. As the first generation of Black men to experience carceral power at home and in prison, Black gangs of the 1960s helped to circulate the performances of masculinity that shaped prison. Because they were formed within the milieu of prison, Black gangs borrowed the hegemonic masculine elements of prison masculinity to form their expressions of masculinity. As "key institutional sites for the expression and reproduction for hegemonic masculinity," prison was a kind of "resource" that enabled them to access masculine authority (Sabo *et al.*, 2001: 6). This appropriation was in keeping with the generational predisposition toward appropriation, reuse, and remixing emblematic of those who grew up affected by economic divestment and deindustrialization that ravaged Black Chicago during the 1970s. The masculinity of Black gangs in turn informed the masculinities of prison. This form of carceral circularity created the context for prisonized gender performances.

This chapter documents the shift in performances of masculinity in Black Chicago between the 1970s and 1990s.[1] It illuminates the shifting performances of carceral masculinity within Illinois state prisons and South Side neighborhoods.

To do this I look at the Almighty Black P. Store Rangers (also called the Blackstone or Stones), a South Side gang that emerged in the 1960s. I seek to provide readers with a "usable past" of carceral Chicago that is formed through genealogy. This chapter is not intended as a history of the Blackstone Rangers but rather, after Foucault, it traces the emergence of a phenomenon. What follows is not a linear history that uncovers "origins" but a "diagnosis" that seeks to re-establish the various system of subjugation that helped give rise to the Blackstone Rangers. To do this I uncover how carceral power in the form of de jure and de facto segregation and targeted policing of South Side neighborhoods gave rise to the Blackstone Rangers and shaped the shifting gender performances of Black men in Illinois prisons and South Side neighborhoods.

This research is situated within a number of scholarly fields; it sits at the intersection of carceral geography, Black studies, and gender studies. This choice is not arbitrary. These scholarly fields illuminate the overlap between spaces of detention, race, and gender. I draw upon these bodies of knowledge to understand, study, and document how anti-Black racism and carceral space informed Black men's performance of gender in Black Chicago.

Prisonizing Black Chicago

Beyond high rates of incarceration, Black Chicago has a long history with carceral power. Chicago's South Side was confronted daily with forms of prison or carceral power that effectively *prisonized the landscape*. By prisonize I mean that techniques and technologies of prison punishment – policing, containment, surveillance, and the establishment of territory, the creation of frontiers – functioned in the quotidian space of Black Chicago. Its primary use has been to "spatialize blackness" in Chicago (Shabazz, 2015).

Carceral power first entered the Black Belt in Chicago vis-à-vis attempts to control interracial sex and socializing in the underground nightclubs in the Black Belt – a seven-mile-long, one-mile-wide strip of land on the South Side. And in doing so it became a permanent fixture. Against the backdrop of the first wave of the Great Black Migration, law enforcement officials, with the support of the Black middle class, began to crack down on the established spaces of interracial sex and socializing. Progressive reformers, with the support of the Black middle class, tried to stamp out vice in the Black Belt. Using new police powers and tactics, and with a larger, more specialized police force, the city of Chicago waged a campaign against vice.

In the early part of the twentieth century, Black/White interracial sexual relations carved out spaces of libidinal pleasure within the Black Belt. Forced there after successful campaigns to close down vice in the segregated Levee district, these spaces of Black/White pleasure, what Kevin Mumford (1997) terms "interzones," operated outside the larger legal, racial, and sexual framework that guided the early twentieth-century racial and sexual politics. Against the backdrop of the Mann Act,[2] which effectively made interracial socializing, sex, and marriage illegal, the interzones changed the nature of sex work in Chicago by creating an underworld for them to take place. Progressive social reformers significantly influenced the political debates about interzones. Their response to vice was influenced by the increasing demographic changes taking place in the city and the implications these changes held for vice districts (Mumford, 1997: 20–21). Threatened by the interracial sexual dynamics of vice in the Black Belt and emboldened by the Progressives' reform initiatives, Chicago city officials adopted a stand against the interzones. The center-piece of their campaign was police power. The city expanded its police force, gave it new powers, and created specialized units to go after vice in the Black Belt. The major problem this produced was that they never left. It became a permanent fixture in the Black Belt.

Once it gained a foothold in the Black Belt, carceral power metastasized, moving across scales and into different parts of Black Chicagoans' lives. Carceral power structured the one-room homes in which Black migrants were forced to live – kitchenettes. These small rooms, inhabited by as many as six people, were the dominant form of housing in the Black Belt because racist restrictions cut off the vast majority of the housing market. The Black writer and social critic Richard Wright (1940) declared that kitchenettes were Black peoples' "prison," their "death sentence without trial." High rates of poverty among Black migrants coupled with housing restrictions and the "unbearable closeness of association" they produced were extensions, indeed the scaling down of carceral power from the community to the home.

In the years following the Second World War, the federal government made a promise of better housing to Black residents in cities like Chicago. After a four-decade "kitchenette" period, the federal government used legislation to change the politics of race and housing in America's cities. The result of that promise was the creation of large-scale public housing. Chicago built massive housing projects on the south and west sides of the city. They were cleaner and roomier than the cramped kitchenettes of the inter-war years; they were also modern and affordable. Inspired by Western European planners, Chicago built tens of thousands of units between 1950 and 1970. This housing not only accommodated existing residents, but it also absorbed migrants from the second leg of the Great Black Migration, who descended onto the city in the years following the Second World War in search of work and to escape southern racism.

However, these massive structures ultimately failed. The demise of the post-war economy drove project residents into poverty. City planning that located the structures in the heart of the segregated Black Belt exacerbated this. High-rise architecture cut residents off from the rest of the community, spatially isolating them in vertical towers. The ubiquity of the police, security cameras, and cages said to protect the residents from the growing instability produced by poverty, imprisoned the population of the projects. This made housing projects in Chicago into a carceral interstice – a place between prison and home. The Blackstone Rangers emerged against this backdrop. Architectures of confinement, restrictions on Black movement, ubiquitous policing, few job options, and rising rates of Black male incarceration provided a fertile carceral landscape for the germination of Black gangs.

Black gangs and carceral space

Chicago's Black gangs date back to the years following the Second World War, coming into existence at the end of a "policy" or gambling era (Hagedorn, 2006: 199). At that point, most of the gangs were unorganized and small. They were populated by young men and often referred to as "clubs" (Moore and Williams, 2011: 18). On the South Side, Black gangs emerged from the neighborhoods in the heart of the Black Belt (Hagedorn, 2006: 200). The Deacons, who started in the early 1940s, claimed the newly built Ida B. Wells housing project as their territory. Nearly two decades later and a mile-and-a-half away, the Blackstone Rangers were born in Woodlawn. The Blackstones emerged amidst the construction of the projects along the State Street corridor, forming in the early 1960s. The early Stones were made up of boys from the Woodlawn neighborhood – a mixed-income, mostly Black neighborhood. Their name came from Blackstone Street, which ran through Woodlawn. At the helm were Eugene "Bull" Hairston and Jeff "Angel" Fort.

Prior to the 1950s, Woodlawn was mostly White. Post-Second World War Black migration from the South changed its racial dynamic, spawning panic among Whites who ultimately fled the Woodlawn area as an increasing number of Blacks moved in. Woodlawn was part of Chicago's expanded Black Belt during the second phase of the Great Black Migration. This second leg of migration was very different from the first; it was made up of mostly poor rural Black people who lacked the kind of sophistication and educational training that had characterized the first wave of migration (Moore and Williams, 2011: 9). Second-wave migrants experienced increased poverty as well as alienation from the first wave of migrants who came in the early part of the twentieth century. This gave rise to profound disillusionment among newcomers (Moore and Williams, 2011: 9–10). Jeff Fort, the co-founder of the P. Stones, was part of the second migration, moving to the city with his mother in the early part of the 1950s. The two-tier system of social and economic organization that defined Woodlawn was the context within which Fort lived.

Black in-migration and White out-migration not only changed Woodlawn's demography but also ushered in another form of the politics of containment. Restrictions on housing in Hyde Park to the north and White communities to the south and around Lake Michigan to the east kept Blacks out of White neighborhoods and trapped in the Black Belt, which by the 1950s had extended further south to 60th Street. For example, when Woodlawn became a mostly Black neighborhood by the mid-1950s, the Chicago Transit Authority (CTA) curtailed its bus service to the area along with many other low-income Black neighborhoods. This had a significant impact upon Woodlawn because it choked off access to the vibrant 63rd Street shopping area, the heart of the neighborhood's economic base (Chepesiuk, 2007: 126). Housing prices plummeted, as did rental prices; slumlords moved in and responsible landowners moved out.

Almost overnight, Woodlawn went from community to ghetto. Ghettoization also meant that the population density increased. As Woodlawn transitioned from White to Black, its population doubled. Woodlawn housing accommodated about 40,000 White people; when Blacks moved in, 80,000 were squeezed into that same housing, representing a windfall to profit-seeking landlords who converted more single-family homes into kitchenettes (Chepesiuk, 2007: 129). When Jeff Fort moved to Woodlawn they lived in a cramped kitchenette (Moore and Williams, 2011: 10).

The walled-in community of Woodlawn was a fertile recruiting ground for the Blackstones. Membership in the Blackstones was an enticing offer to the thousands of young men and boys who spilled onto the streets; many of them had too much time on their hands and too few job prospects. The Stones thumbed their noses at authority; they were brash, stylish, and flush with cash from Great Society programs and independent funders (Siddon, 1970). During the 1960s, the federal government under the Johnson administration, along with the Democrats, created a series of programs and initiatives to address urban poverty and racial inequality. The Woodlawn Organization, which was started in the early 1960s and led by Nicholas von Hoffman, a young White organizer, received federal funds to work with the Stones. The hope was that if the Stones could be drawn in and politicized, this would lessen the friction between the Stones and their chief rivals, the Disciples. The Stones accepted the partnership but used this mainstream support to strengthen and expand their organization, which caused deep resentment from the Disciples, who lacked institutional support (Moore and Williams, 2011: 14, 50, 52).

Public space and Black masculinity

Ask any Black Chicagoan about the Stones and you will likely hear one of two responses. Some will say that the Stones were a street organization which tried to exact control over Black communities in the face of ongoing economic and political divestment on the part of the White establishment; others will surely say that they were thugs who exploited the Black community for their own selfish needs. There is no singular way to characterize the Blackstones; they were many things at different times. More than revolutionaries or thugs – or a combination of both – the Blackstone Rangers were also "architects of social space" in a setting that was often unkind to them (Davis, 1992: 293). Amidst segregation, restrictions on housing, and rising overpopulation, the Blackstones made it possible for young Black boys like Fort – who felt alienated from their new urban settings – to construct identities and geographic belonging. Being a Stone gave Black male youths something to which they could lay claim. Fort recognized that without a place, the Blackstone identity could not endure. With so little space in Woodlawn, Fort used all his know-how to find them "a place of their own"; whether a church, pool hall, corner, apartment building, or "makeshift" temple, place was vital to the Blackstone ethos (Moore and Williams, 2011: 28–29).

Being a Stone was a way to contest the complicated relationship Black men have with public space. Stripped of what Craig Wilkins terms "spatial empowerment," Black men have fought for access to public places both within and outside Black geographies (2007: 21). Considered a public nuisance and a threat, Black men have been and remain targets of police power. Black men are cast as the nation's "bogeymen" – the racial embodiment of terror. As a result they are confronted with forms of racist violence and regulation in public space. Police power plays a significant role in this. Recent police killings of unarmed Black men in Ferguson, Missouri; Cleveland, Ohio; Brooklyn, New York; Staten Island, New York; and Beavercreek, Ohio are illustrations of the state-sanctioned violence Black men endure daily in public space.

In the face of this violence or claims that they are "out of place," Black men contested their spatial disempowerment in mid-twentieth-century Chicago. One of the ways they did this was through inhabiting public spaces and using them as sites to create and perform gender. As feminist geographers have demonstrated, public space is deeply gendered.[3] Black communities are the places Black men use to contest violence and exclusion. For many Black men, Black communities are the center of their spatial mobility, identity production, and resistance (Tyner, 2006: 62). Black men are not foreign to or out of place in Black communities. In the face of police harassment from within Black communities, Black men nevertheless reflect the stereotyped norm (Wilkins, 2007). White men, on the other hand, can access public space across locations, giving them (and their gendered masculine performances) the ability to inhabit these spaces without opposition.

Black men's quest for spaces to be men has not always been productive. In some cases their desire to lay claim to public spaces in an attempt to access masculinity and resist racial and gender marginalization has produced over-compensatory and exclusionary actions. The geography of hip-hop is an example. Seen as threats to the public, Black women and men used the geography of Black communities to create spaces for the cultivation and expression of hip-hop. However, Black men "sought to control the new geographies of resistance by attempting to exclude women" and sexual marginals (Shabazz, 2014: 4). This "contradictory openness through exclusion" also marked the Blackstones' creation of social space (Shabazz, 2013: 4).

Despite trying to create a positive identity and sense of belonging, the social space the Stones fostered was essentially rooted in chaos. Because the social space they created was constituted within the politics of racist containment, tension, and ultimately violence, these elements became part of their social world. Through an attempt to access the public space of Black Chicago, young men like Jeff Fort and Bull Harrison forged alliances to not only inhabit the space of the community, but on some level to control it. Given these conditions, according to Hagedorn, "it should be no surprise that" Woodlawn produced "angry and alienated groups of armed young men" (2008: 53). And as other young Black males formed similar groups under the same forbidding racist circumstances, these Black urban simulacrums lethally clashed. Indeed, against the backdrop of containment, poverty, joblessness, racist policing, and the state-sponsored destruction of Black radical movements, what bell hooks calls the "slaughter of radical black men," the Stones grew stronger (Hagedorn, 2008: 57). Under these conditions the Stones were able to come together as a collective force to fend off challenges from other groups and even thrive under difficult socio-spatial conditions. This had complex and contradictory outcomes. On the one hand, thriving in Woodlawn gave the Stones access to the community – in particular the public space – which non-gang-affiliated young men did not have. Existing as a part of a collective allowed the Stones to fend off bullies or challenges from other groups. But creating this dynamic required the use of lethal force, particularly against other Black men. As architects of social space, the Stones, like most gangs, tried to control and defend specific territory (Hagedorn, 2006: 91; Zilberg, 2011; Peralta, 2009), carving up neighborhoods and claiming certain spaces under their own banner.

Stone masculinity

Many forms of masculinity existed among the Blackstones, but toughness was a dominant current running through it. Against the backdrop of unattainable forms of masculinity that Black boys and men could not access within the context of poverty, joblessness, and spatial containment, toughness was an accessible form of masculinity that created forms of respect and recognition.

The original Stones emerged from a pool of alienated, frustrated, idle young men who hung out on the Woodlawn streets between 63rd and 64th and Blackstone. Fort was the leader of the gang. They often got into trouble. They fought other boys; they were accused of stealing from local stores and snatching purses from women in the community. When they met up with "Bull" Hariston and his crew, they joined forces and the Blackstone Rangers were born. It may be argued that these young boys were simply bad seeds. Some who knew Jeff Fort saw him this way, arguing that even from a young age he had demonstrated a tendency toward the nefarious. I do not disagree with this position; rather, I argue that the socio-spatial conditions of Woodlawn, compounded with an inability to access patriarchal power, played a significant role in shaping the lives of these troubled young men. In an overcrowded, contained, and deeply impoverished world where most people lacked resources and the ability to exercise meaningful power over their lives, where joblessness was high and Black men could not "provide" in the patriarchal sense, personal articulations of agency in the form of toughness, fighting, working in the underground economy, or stealing from others provided access to resources and generated prestige and recognition as well as symbolic forms of power. This is not to say that Jeff Fort and his crew were not opportunists who exploited vulnerable people. But there was more to them than this. Fort responded to his environmental and gendered circumstances by performing deeply problematic notions of masculinity. However, Fort was also able to use the persona he created to protect himself from a world that was hostile and unkind.

This gender performance was a consequence of containment and poverty, but these were not the only factors involved. While containment and poverty raised tensions, Black men's "patriarchal socialization," which tells them that they must get a job and provide for and protect their families, did much to move tension into violence (hooks, 2000: 85). It does this because of the paradox within which Black masculinity is locked: while boys are socialized under this form of masculinity, most Black men cannot access it (Majors and Billson, 1992; hooks, 2000; Estes, 2005; Belton, 1995). The inability to access these culturally dominant narratives of masculinity against the backdrop of poverty and containment can make seemingly mundane objects like control of a street corner or a sign of disrespect cause for physical altercation. Black men's bodies play a key role in this performance of masculinity. As the "original subject that constitutes spaces," to use the philosopher Merleau-Ponty's words, the body is the site where Black men perform masculinity (quoted in Whitehead, 2002: 185). The body, therefore, stands in for or compensates for patriarchal power. In the social space of Woodlawn, amidst geographic restrictions on housing, overcrowding, poverty, and rising crime rates, embodied performances of masculinity in the form of a tough exterior, macho performance, willingness to use force, or the domination of other men, predicated these performances of masculinity. Masculinity scholars see this tough exterior as a kind of "coping strategy" (Majors and Billson, 1992: 28). This strategy made it possible for Black men in Woodlawn to access a characteristic of patriarchy that did not require material forms of power: toughness.

Toughness became an organizing logic for the Blackstone Rangers. While not every member saw masculinity the same way, they nevertheless understood that toughness and a willingness to do harm to others if provoked were normative displays of Stone manhood. This performance of masculinity ultimately served to benefit the growth of the organization. Like the military, which uses culturally dominant ideas of masculinity – some of which are institutionalized as a means of filling the ranks (e.g., all boys in the United States must register with the Selective Service when they turn 18) – the Blackstone Rangers were able to grow and become a strong organization through normalizing tough-guy masculinity.

By the latter part of the 1960s, the war over street territory between the Blackstone Rangers and the Disciples had spilled much blood on the streets of Woodlawn and Englewood, a mostly poor Black community east of Woodlawn. As a result the police began to focus their attention on Black gangs. In 1967, the Chicago Police Department created the Gang Intelligence Unit (GIU), the first organized response to Black street gangs in Chicago. The purpose of this unit was to centralize the decentralized response to gangs, and force was their primary tool. The GIU raided Stone hangouts, confiscated weapons, examined their finances, and made arrests. The state and federal government indicted members and targeted leadership in an effort to bring down the organization (O'Brien, 1971; Jones, 1970; Lee, 1970a, 1970b; Black P Stone Leader Seized, 1970). The unit felt justified in using these tactics in part because residents of Woodlawn urgently sought a reprieve from gang violence. Despite their tough tactics, the GIU were never able to ease the violence between gangs. What they did accomplish was to re-spatialize gang tensions from the street to correctional institutions. By the latter part of the 1960s and the beginning of the 1970s, the GIU had a hand in incarcerating numerous members of gangs as well as leaders including Bull Harris and Jeff Fort.

P Stones and prison masculinity

While the geography of containment provided the context for the emergence of the Stones and the gender performance rituals that informed the group, state incarceration altered again their performances of masculinity. With the aid of the GIU and the willingness of the State of Illinois to spend resources to incarcerate gang members, by the early 1970s Black gangs were a growing population in the state's prisons (Jacobs, 1977). Jeff Fort and Bull Harrison were part of the wave of gang members incarcerated during this period. When they entered Stateville Penitentiary in Crest Hill, Illinois, hundreds of other gang members were awaiting them. The preoccupation with the rise of crime in the city government and state legislature buoyed the expansion of policing within parts of the Black Belt, swelling the state prison population. Prisons had a profound impact upon gangs, because they enabled gangs to recruit in much more effective ways than they could on the outside. Both the Blackstone Rangers and the Disciples used prison to boost their membership (Moore and Williams, 2011). They were able to do this because the conditions of prison – the violence in particular – forced men to find avenues for safety. Gangs were key in this because membership could insulate individual prisoners from violence. But more than a fruitful recruiting ground, prisons provided gangs with new scripts for the performance of masculinity.

Masculinity in prison in the United States – "prison masculinity" – is a complex, contradictory performance of masculinity within carceral institutions such as jails, prisons, and juvenile facilities (Sabo *et al.*, 2001). The single-sex environment and the mostly male prison staff "represents on the whole a more intense version of gender, or more accurately, of masculinity" (Seymour, 2003: 33). The intensity of gender performance in prison reinforces the gendered dynamic of prison. Prison masculinity, therefore, is a performance of masculinity that circulates within the brutal and hegemonic carceral setting. It is a function of the patriarchal logic of prison. In the United States, decades of congressional calls for harsher sentences (spurred by the "war on drugs") made prisons into institutions where hegemonic masculinity reigned supreme. Prisons are not patriarchal by nature, but rather were made that way through a series of policy moves. The patriarchy of the institution is demonstrated in its homo-sociality, sex segregation, hierarchy, and violence (Sabo *et al.*, 2001: 7–8). As a result, prisons in the US and elsewhere have become places where the will to hurt others, the celebration of toughness, subjugation, domination, hierarchy, and social control are part of the gender landscape (Martin and Jurik, 1996: 171; Seymour, 2003: 34; Sabo *et al.*, 2001). Moreover, and in spite of the closeted nature of prisons, the masculinities within them are not created in vacuums; they are an amalgamation of several masculinities that exist both on the inside and the outside. The term "prison masculinity" is therefore an oversimplification. The fact is that a range of masculinities that come from multiple institutions, geographic locations, classes, sexualities, and racial backgrounds shapes masculinity in prison. Prisons are not "isolated institutions" where masculinities are formed (Sabo *et al.*, 2001: 5); rather they are more like a gender hub where masculinities collide, mix, and mesh.

Black men in the United States have a unique relationship with these masculinities. The geographically specific forms of subjugation that have shaped Black Americans (slavery, Jim Crow, ghettoization, segregation) created the context for the rise of Black mass incarnation and the prisonization of Black masculinity. As I have written elsewhere (Shabazz, 2013), carceral power and the spatial forms it engenders have played an important role in the contemporary performances of Black masculinity. In Black Chicago, for example, the targeted policing of Black communities created through de jure and de facto segregation, and the revolving door between prison and home, has exposed large numbers of Black men to prison masculinities. This has resulted in prison masculinity becoming part of Black men's gender performance in ways that White men are not forced to confront. This racialized form of prison masculinity, generated through the particular social formation of anti-Black racism, is a fact to which the international literature on prison masculinities has not been attentive (see Shabazz, 2009, 2015).

The grammar of man

Violence, subjugation, domination, and toughness are not the only elements of prison masculinity in the United States. Love, friendship, loyalty, and intellectual engagement are also part of masculinity in carceral institutions (Jackson, 1994; Abu-Jamal, 1995, 1997, 2004), as is personal transformation. For many Black men in prison, religious conversion is a way for them to change their lives. For the past half-century, Black men have experienced a spiritual awakening in prison. Islam has had the largest impact in terms of conversions while in prison (Haley, 1964; Leonard, 2003). Black people have deep connections with organized religion (mostly Christianity). As a result, many have and continue to use it as a pathway out of despair. Moreover, given the awful conditions of life in prison, religion offers hope and the possibility of redemption, a point illustrated in the letters of countless fathers, sons, uncles, brothers, boyfriends, and husbands who have found their way to religion in prison.

Beyond being a way to reform oneself however, Islam has played a more prominent role in shaping the performance of Black masculinity. Malcolm X, for example, reinvented himself in prison, not simply through religious transformation or study, but by adopting the codes, posture, the rules, ideology, and dress of the Nation of Islam. I term this *the grammar of masculinity*. I use "grammar" to communicate how men adopt the set of rules necessary to perform masculinities. By casting off the shell of the hyper-masculine hustler (which had its own distinct grammar) and becoming a disciplined, restrained, articulate Black Muslim, Malcolm became a new man (Marable, 2011).

Jeff Fort entered prison in 1972 and was released in 1977; like Malcolm, he emerged a new man. He looked different. The skinny, shaven man who entered prison exited a bulked-up and bearded one. But more than a physical transformation, prison gave Fort a new way of being a man. While he was inside, Jeff Fort became a member of the Moorish Science Temple of America, an Islamic religious sect dedicated to the spiritual, economic, and social elevation of Black Americans. Islam provided Fort with a new spiritual tool that he found useful in reorganizing the Stones. It put at his disposal an ironclad form of discipline that he used to reformat how the men under his leadership understood and performed masculinity. Men in prison often perform a masculinity that is an exacerbated version of the way they enact masculinity on the outside. The gender hierarchy within prison reinforces acceptable ways of being male. Being a Muslim is one of the established gender performances within prison. This form of Black masculinity, which emphasized discipline, order, dutifulness, forthrightness, respect, and, above all, submission to Allah, became the new Stone masculinity Fort tried to instill in the men under his leadership. This differed significantly from Stone manhood that centered on protecting street territory and toughness.

The masculine performance of Black Muslims became synonymous with discipline and organization. Groups like the Nation of Islam constructed masculinity in large part by demanding that members adhere to a hierarchy, similar to that of the military. They adhered to the guiding principles of no drinking in public, no drug use, and being respectful of Black women; an ideology; uniforms; and daily practices. Permeating this order were performative scripts that spelled out a new vision of manhood. Organization was at the core of this script. This form of prison masculinity is different from the contemporary notions of manhood, in which individual toughness and hyper-masculinity occupy the top spot in the hierarchy. This is not to say that dominant forms of masculinity, which emphasized toughness, violence, and a will to do harm were not part of prison's gender hierarchy during the 1970s; however, these forms of masculine bravado were often eclipsed by performances of masculinity that privileged duty, discipline, and submission to the group.

Fort used Islam to install a top-down organizational structure that demanded discipline and unconditional loyalty. Islam featured prominently in this new organizational form and Fort was able to use the discipline embedded in the religious practice in the service of building the Stones. Fort knew that in order to build the organization he would have to build a new kind of Stone, a new kind of man, influenced by the politics of the Nation of Islam (NOI). The NOI built their organization by transforming Black men who suffered from racism into respected, disciplined men. Fort hoped to mirror these efforts. Following the NOI playbook, he imposed a tightly organized structure upon the men around him. He called them the El Rukns, an essentially secret society of men, handpicked by Fort. El Rukn is an Arabic term meaning cornerstone of the Islamic holy site Kaaba, which is the large black rock Muslims circle seven times during the Hajj (Moore and Williams, 2011: 132). Fort used them to not only carry out prison and street activities such as selling drugs and in some cases committing murder, but also to model for the rest of the organization what the new Stone masculinity looked like (O'Brien, 1972).

The El Rukns were a central part of the performative script of 1970s prison masculinity. In Stateville Prison, which was thick with Blackstones, the tight organization Fort built became the syntax of prison masculinity, mandating obedience, discipline, truthfulness, and submission. These traits were mixed with the machine-style political organization. As they had done a decade before on the streets of Woodlawn, the Stones were remaking masculinity. And just as before, they were doing it within the context of containment, which helped the Blackstone Rangers recruit members in prison. Although Black Muslim sects like the Nation of Islam grew in Illinois prisons, it was Black gangs that bene-fited the most. Throughout the 1970s and 1980s, the prison system proved fertile recruiting ground for gangs (Venkatesh, 2000; Jacobs, 1977).

Prison masculinity for a new era

Fort's strategies found a foothold in prison in the 1970s. There, he was able to create a small cadre of loyalists. However, installing this new form of Islam into the P. Stone Nation with men who had spent their lives as Blackstone Rangers on the outside proved a difficult task. When Fort was released from prison in 1977 a crowd of Stones cheered him, eager to see their leader. Nevertheless, his disciples found Islam a hard pill to swallow. Religion constituted part of the tension between Fort (now Chief Malic) and older Stones, because they saw themselves as Christians, not Muslims. The biggest tension was the gender per-formance within Islam, which emphasized duty, submission, collectivity, and discipline. This performance contradicted the male persona fostered among 1960s gangsters. Shifting ideals of masculinity, largely couched in terms of ideo-logy and leadership, produced tensions and splintered the organization.

The tension between the El Rukns and the older Stones mirrored the genera-tional shift between prison masculinity of the 1970s and prison masculinity in the period after 1980 as well. Fort's need for a greater sense of belonging, order, discipline, and ideology produced a performance of prison masculinity that was religious in nature. This performance did not remain the dominant form; it was supplanted by performances that reflected the social forces bringing a new gen-eration of Black men to prison amidst a transformation in public policy. This new cohort needed a different performance of masculinity, one that reflected the changing function of prison. Unlike Jeff Fort, these men entered prison against the backdrop of the Draconian war on drugs, the most devastating form of social policy since slavery.

Between the early 1970s and the mid-2000s, the United States declared and waged war against poor and working-class people of color, continuing the discur-sive construction of them as irrevocably doomed to crime while also producing the social relations that criminalize them (James, 2007; Gilmore, 2007; Davis, 2003; Wacquant, 2009; Alexander, 2010). Although President Nixon coined the term in 1971, the war against drugs is actually a century old. In that time, many lives have been unnecessarily lost to what one scholar calls America's crusade our "Holy War" (Benavie, 2008). Between 1914, when the first piece of anti-drug legislation was passed, until the early 1970s, illicit drugs shifted from being a benign source of personal indulgence into a major political topic that captivated the attention of the nation, spurring the creation of a mega-industry built on pure punishment.

Nixon's declaration rapidly accelerated the diversion of public funds into the punishment industry. On the heels of the 1968 Omnibus Crime Control and Safe Streets Act, the Nixon administration passed the 1970 Comprehensive Drug Abuse and Prevention Control Act. This legislation set off a chain reaction that used the power of the state to punish poor Blacks in urban centers around the nation. Several other federal laws in the same vein followed, including the 1984 Comprehensive Crime Control Act, which established mandatory minimum sentencing for carrying a gun during drug incidents or violent crimes; the sentencing Reform Act which effectively phased out the federal parole system; the Anti-Drug Abuse Act of 1986 that allowed Congress to establish mandatory minimum sentencing for illegal drug use; the 1988 Omnibus Anti-Drug Abuse Act that created mandatory minimum sentences for possession and distribution of crack cocaine – while powdered cocaine remained a misdemeanor; and the 1994 Federal Crime Bill which required states to mandate that 85 percent of a prisoner's term be served in order for states to qualify for anti-crime federal funding. This list does not include the numerous state legislative acts that went on the books, like the Rockefeller Drug Laws of 1973 or the 1994 Three-Strikes laws in California, which stiffened already stringent federal laws, not to mention the copious judicial acts which greatly expanded the power of state prosecutors and judges to inflict more punishment on those swept up in ravenous anti-drug crusades. All the funding and the judicial acts and sentencing laws were enabled in large part by the explosion of police power.

Policing during the early 1980s radically transformed the tone and tenor of the drug war. Under the Reagan administration, the police gained unprecedented powers that they used for waging war. This fact is demonstrated most ardently by drug convictions. Since 1980, the number of people convicted on drug charges has increased by 1,100 percent (Alexander, 2010). Those incarcerated on drug charges account for a significant number – more than half between 1980 and 2000 (Alexander, 2010). More than 31 million people have been arrested since the start of the drug war (Mauer, 2006). At the beginning of the current crusade in 1972, the Supreme Court granted the police powers that undermine constitutional protections, such as the right not to be subject to illegal search and seizure. Police are now able to utilize race as criteria for stopping citizens to search for drugs, while federal and state funds have enabled an unprecedented militarization of police forces.

Men who entered prison during the drug war era had a different experience with the tropes of masculinity than they experienced behind bars. By the mid- to late 1980s the model of masculinity-as-organizational tactic faded. The new prison masculinity extolled the virtues of individual toughness, personal strength, the will to hurt others (if necessary), and individual over collective strength. Moreover, the landscape – that is to say, the place in which this new form of masculinity was expressed – also changed. For this new generation of incarcerated men, the body was the site for the expression of masculinity. In contrast to those markers of masculinity in the 1970s, large muscles, hard stares, and imposing postures became the hallmarks of prison manhood. The body was not just the site of expression for this generation of prisoners; it was the mechanism through which they spoke.

What facilitated the change from prison-masculinity-as-organization to one of toughness expressed through a single body? The police crackdown, spearheaded by the GIU, brought the P. Stone Rangers and other Black gangs in Chicago into prison during the latter part of the 1960s and the early 1970s; it was a crackdown foregrounded in large part against the backdrop of politics. The P. Stones saw themselves as part of a larger political reality. They identified with Black Nationalism and even attempted to forge an alliance with the Black Panther Party (Hagedorn, 2006: 201; Moore and Williams, 2011: 96–99). As these gangs came to prison this political context followed them. However, the next generation of Black men who entered prison in the latter part of the 1970s and early part of the 1980s did not have this kind of political framework. For these prisoners, there was no political organizing in prison and no Attica. Even the prison social system which emphasized collective organizing among gang members began to loosen and fray, and there was no organized structure in place to latch on to. This had two distinct consequences: on the one hand, this absence disrupted the organizing efforts prisoners created in the early and mid-1970s. On the other hand, it decentered organization as the hegemonic form of masculinity, creating a vacuum at the top of the prison gender hierarchy.

Concluding reflections

As this chapter has demonstrated, carceral power and public policy played an active role in the creation of Black gangs and their performances of Black masculinity on Chicago's South Side. This research begs the question: What can be done to change this? How can these "sores in the city" be healed? I offer three suggestions. First, Black communities need fewer police. The ubiquitous presence of the police in Black communities does not produce safety. On the contrary, for many residents the police are agents of fear and frustration. Fewer police in Black communities would mean less harassment and profiling, and fewer arrests and deaths of Black people. The money saved by scaling down policing may be used to fund and expand the work of violence interrupters like Ceasefire.

Second, the city of Chicago must invest in communities like Woodlawn. The legacy of institutional segregation which produced the Black Belt and the gangs that followed undermined the economic and social stability of this community. Chicago must not look to gentrification (i.e., Black removal) as the response to poverty and instability. The city must address this wrong by using its power to inject resources into this community. Investment will give community members the ability to deal with street-level crime and confront the gang issue. Gangs thrive in places with instability.

Third, the state must rethink its four-decade-long march toward mass incarceration. The war on drugs, which has been fought in Black Chicago, has devastated communities, wasted resources, broken up families, and created generations of young men who are unprepared for work, school, and civic engagement. The resources spent on incarceration can rebuild communities impacted by economic downturns all over the state. Mass incarceration has been an utter failure, and it is time that the Illinois legislature and the political leadership in the city of Chicago realize this.

Healing the sores in the city also demands that the usable past this scholarship raises is taken seriously. This work on the Blackstone Rangers in Chicago illustrates the extent to which the past lives in the present. My genealogy of the Stones puts under a microscope the ways in which Chicago's socio-spatial landscape ushered in the rise of Black gangs and the gender performances that accompanied them both inside and outside prisons. The Stones teach us that carceral power not only produces forms of punishment; it also produces subjects.

Notes

* This is a revised chapter from *Spatializing Blackness: Architectures of Confinement and Black Masculinity in Chicago* by Rashad Shabazz. Copyright 2015 by the Board of Trustees of the University of Illinois. Used with permission of the University of Illinois Press.
1 While I focus on the period between 1970 and 1990, I situate my historical geography in a broader context that extends from the early part of the twentieth century to the dawn of the twenty-first century. Through an examination of Black migration to Chicago from the years after the First World War to the rise of Black gangs in the postwar period, I delineate how the spatialization of blackness on Chicago's South Side set in motion the processes that enabled the prisonization of Black Chicago and the shifts in performances of Black masculinity. Framing my historical analysis in this way makes it possible to draw attention to the shift in Black masculinity that took place between the 1970s and 1990s.
2 The Mann Act, in no uncertain terms, demonstrated the extent to which the state would go to limit interracial socializing. Congress attempted to regulate prostitution through making the transportation of prostitutes across state lines illegal for the purpose of "immoral sexual relations." Using federal power to drive a wedge between interracial sex and marriage made it possible for states to take steps to do the same (Mumford, 1997: 10–12).
3 The feminist scholarship that informs my thinking focuses its attention on White, mostly middle-class women. Therefore, I am cognizant of the limitations of this research for drawing conclusions about Black women and men and access to public space. Racism has and continues to play an important role in marginalizing Black women's and men's access to space. Yet the analysis of feminist geographers such as McDowell (1999) and Wilson (1991) is, nevertheless, helpful in shading in the borders that define the uneven distribution of geographic power. They help me to understand the broader struggle of women and men across race, class, gender, and sexuality with respect to social realities that shape access to social space. Therefore, I accept their assessment that there is a gendered division of spatial access while recognizing that it is mediated through social forces like race. For understanding masculinities specifically, see Whitehead (2002), Kimmel (2005), and Connell (2005).

References

Abu-Jamal M (1995) *Live from Death Row.* Reading, MA: Reading Addison-Wesley.
Abu-Jamal M (1997) *Death Blossoms: Reflections from a Prisoner of Conscience.* Farmington, PA: Plough Publishing.

Abu-Jamal M (2004) *We Want Freedom: A Life in the Black Panther Party.* Cambridge, MA: South End Press.

Alexander M (2010) *The New Jim Crow.* New York: The New Press.

Belton D (ed) (1995) *Speak My Name: Black Men on Masculinity and the American Dream.* Boston, MA: Beacon Press.

Benavie A (2008) *Drugs: America's Holy War.* New York: Routledge.

Black P Stone Leader Seized in Hotel Raid (1970) *Chicago Tribune*, July 27.

Brooks G (1987) *Blacks.* Chicago, IL: Third World Press.

Chepesiuk R (2007) *Black Gangsters of Chicago.* Fort Lee, NJ: Barricade Books.

Chesney-Lind M and Mauer M (2003) *Invisible Punishment: The Collateral Consequences of Mass Imprisonment.* New York: New Press.

Connell RW (2005) *Masculinities.* Berkeley: University of California Press.

Davis AY (2003) *Are Prisons Obsolete?* New York: Seven Stories.

Davis M (1992) *City of Quartz: Excavating the Future in Los Angeles.* New York: Vintage Books.

Dawley D (1992) *A Nation of Lords: The Autobiography of the Vice Lords.* Illinois: Prospect Heights.

Drake SCAC and Horace R (1945) *Black Metropolis: A Study of Negro Life in a Northern City.* Chicago, IL: University of Chicago Press.

Estes S (2005) *I am a Man: Race Manhood, And The Civil Rights Movement.* Chapel Hill: University of North Carolina Press.

Gilmore RW (2007) *Golden Gulag: Prisons, Surplus, Crisis, and Opposition in Globalizing California.* Berkeley: University of California Press.

Hagedorn JM (2006) Race not space: A revisionist history of gangs in Chicago. *The Journal of African American History* 91: 194–208.

Hagedorn JM (2008) *A World of Gangs.* Minneapolis: University of Minnesota Press.

Haley MXWA (1964) *The Autobiography of Malcolm X.* New York: Ballantine.

Hill-Collins P (2004) *Black Sexual Politics.* New York: Routledge.

hooks b (2000) *We Real Cool.* New York: Routledge.

Jackson G (1994) *Soledad Brother: The Prison Letters of George Jackson.* Chicago, IL: Lawrence Hill Books.

Jacobs JB (1977) *Stateville: The Penitentiary in Mass Society.* Chicago, IL: University of Chicago Press.

James J (ed) (2007) *Warfare in the American Homeland.* Durham, NC: Duke University Press.

Jones W (1970) U.S. to probe how Stones spent grant. *Chicago Tribune*, September 10.

Kimmel MS (2005) *The History of Men: Essays in the History of American and British Masculinities.* Albany: State University of New York Press.

Lee E (1970a) 7 Stones indicted in cop death. *Chicago Tribune*, September 4.

Lee E (1970b) Jurors indict Fort, 17 other gang members. *Chicago Tribune*, January 30.

Leonard KI (2003) *Muslims in the United States.* New York: Russell Sage Foundation.

Lipsitz G (2011) *How Racism Takes Place.* Philadelphia, PA: Temple University Press.

Majors R and Billson JM (1992) *Cool Pose: The Dilemmas of Black Manhood in America.* New York: Macmillan.

Marable M (2011) *Malcolm X: A Life of Reinvention.* New York: Viking.

Martin SE and Jurik N (2006) *Doing Justice, Doing Gender: Women in Legal and Criminal Justice Occupations.* Thousand Oaks, CA: Sage.

Mauer M (2006) *Race to Incarcerate.* New York: The New Press.

McDowell L (1999) *Gender, Identity and Place: Understanding Feminist Geographies.* Cambridge: Polity Press.

McKittrick K (2006) *Demonic Grounds: Black Women and the Cartographies of Struggle.* Minneapolis: University of Minnesota Press.

Moore NY and Williams L (2011) *The Almighty Black P Stone Nation.* Chicago, IL: Lawrence Hill Books.

Mumford KJ (1997) *Interzones: Black/White Sex Districts in Chicago and New York in the Early Twentieth Century.* New York: Columbia University Press.

Mutua AD (ed) (2006) *Progressive Black Masculinities.* New York: Routledge.

Nast HJ and Pile S (1998) *Places Through the Body.* London: Routledge.

O'Brien J (1971) Police raid Stones' arsenal, arrest 7. *Chicago Tribune*, February 13.

O'Brien J (1972) Black P Stone Nation muscling into drug racket, police say. *Chicago Tribune*, June 11.

Peralta S (2009) *Crips and Bloods: Made in America.* Produced by The Gang Documentary, Balance Vector Productions and Verso Entertainment.

Reeser TW (2010) *Masculinities in Theory.* Malden, MA: Wiley-Blackwell.

Rios V (2009) The consequences of the criminal justice pipeline on Black and Latino masculinity. *The Annals of the American Academy of Political and Social Science* 623: 150–162.

Sabo DF, Kupers TA, and London WJ (eds) (2001) *Prison Masculinities.* Philadelphia, PA: Temple University Press.

Seymour K (2003) Imprisoning masculinity. *Sexuality and Culture* 7: 27–55.

Shabazz R (2009) "So high you can't get over it, so low you can't get under it": Carceral spatiality and Black masculinities in the United States and South Africa. *Souls* 11: 276–294.

Shabazz R (2014) Masculinity and the mic: Confronting the uneven geography of hip-hop. *Gender, Place, and Culture* 21: 370–386.

Shabazz R (2015) *Spatializing Blackness: Architectures of Confinement and Black Masculinity in Chicago.* Urbana: University of Illinois Press.

Siddon A (1970) "I finance Stones" Kettering heir says. *Chicago Tribune*, August 22.

Tyner JA (2006) *The Geography of Malcolm X: Black Radicalism and the Remaking of American Space.* New York: Routledge.

Venkatesh SA (2000) *American Project.* Cambridge, MA: Harvard University Press.

Wacquant L (2009) *Prisons of Poverty.* Minneapolis: University of Minnesotta Press.

Whitehead S (2002) *Men and Masculinities.* Malden, MA: Polity Press.

Wilkins CL (2007) *The Aesthetics of Equity: Notes on Race, Space, Architecture, and Music.* Minneapolis: University of Minnesota Press.

Wilson E (1991) *The Sphinx in the City: Urban Life, the Control of Disorder, and Women.* Berkeley: University of California Press.

Wright R (1940) *Native Son.* New York: Harper Perennial.

Zilberg E (2011) *Space of Detention: The Making of a Transnational Gang Crisis between Los Angeles and San Salvador.* Durham, NC: Duke University Press.

Part II

Prisons as artifacts in historical-cultural transition

5 Doing time-travel

Performing past and present at the prison museum

Jennifer Turner and Kimberley Peters

Introduction

The transformation of former prisons into sites of "dark tourism" reflects a recent trend in the use of decommissioned buildings for alternative purposes, such as museums and other heritage sites, which emphasize "representations of death, disaster or atrocity for pedagogical and commercial purposes" (Walby and Piché, 2011: 452). Prisons are spaces that hold a morbid fascination for visitors who are unlikely to ever encounter such a space in their everyday lives (Strange and Kempa, 2003). Far from being a traditional tourist site, the prison museum is built upon consumer desire to access the inaccessible; to glimpse a life on the "inside" and all its assumed horrors from the comfort of being on the "outside" (Turner, 2013) – with the choice and liberty, to enter, to leave, to accept, or to reject any given exhibition or display (see Hall, 1973). Prison museums cater, on the one hand, to a market of visitors seeking such tourist experiences for entertainment (Adams, 2001; Schrift, 2004). On the other hand, they function to educate visitors about penal pasts, shaping contemporary understandings through engagement with carceral histories (see e.g., Baker, 2014: 1).

In this chapter we focus on the ways in which a particular prison museum – the Galleries of Justice, in Nottingham, UK – informs and entertains while making the past usable in the present. The Galleries of Justice is a prison museum housed within a former courthouse and jail. It not only tells a penal history specific to the county of Nottinghamshire, UK; it also conveys a national carceral history, holding the official *Her Majesty's Prison Service* archive collection. The museum also portrays the global connections of Britain's penal past through exhibitions charting the period of transportation to the Americas and Australia (Baker, 2014; Galleries of Justice, n.d.d). During a two-year period of research investigating the manner in which prison museums allow those on the "outside" to access life "inside" (see Turner, 2013), we found performance to be a crucial technique in making the past usable and comprehensible to the visitor. By performance, we refer to the ways in which the past is not merely represented, but also embodied and brought to life in so-called "non-representational" ways (Thrift, 1996) – through costumed interpreters, audio guides, and by encouraging visitors to participate themselves as pseudo-convicts.

During one of the eight visits we made to the Galleries of Justice, we participated in a group tour of the former prison. As the tour progressed through the courtrooms to the prison cells below, we moved linearly through time while also experiencing overlapping time periods in the same space. In the exercise yard, for example, we experienced 1856 and 2013 simultaneously:

> Before entering the women's prison, we were lined up and inspected in the exercise yard by Mrs Linton, the prison matron. We were shouted at in a stern voice for our slack shoulders and ordered to open our hands out to scrutinise our nails and ensure we were not hiding anything. She then informed us how we must not "cross her" (after all, her husband was the governor); we were to "obey" orders at all times; speak "only when spoken to" and "co-operate" with her and her staff. We were told, in no uncertain terms that we would be "reformed through hard labour and education". Any inclination to titter at the performance was suppressed by our being chastised by the lady dressed as a 19th century matron. She was really quite serious. Had we smiled, we felt we might be punished as described. We followed her direction, marching around the exercise yard when explicitly told to – just like the prisoners of the past.
>
> (Turner and Peters, 2013)

Performances, we argue, are crucial to the ways in which carceral pasts are used in the present. Performance is often used to bridge a temporal gap between past and present at heritage sites (see Leighton, 2007). Yet, in the penal heritage site, such performance becomes doubly significant. It marks a spatial boundary crossing from the outside to life inside – creating a sense of life somewhere distinctly different – a space hitherto unknown to the visitor. Yet performance also works to facilitate a temporal boundary crossing as tourists engage various penal pasts; they embark on a process of time-travel to distant and not-so-distant pasts via the performed narratives constructed at the museum. We argue that this builds an affective experience of past carceral space that also shapes perceptions of prisons in present society.

In this chapter we explore the performative time-travel techniques of the Galleries of Justice in order to interrogate how this museum makes British penal history knowable and usable to the audience. To do so, we work through varying strategies of performance: the performance of museum staff through costumed interpretation and the performed audio tours and virtual tours; and the encouraged performance of visitors to co-produce these historical narratives through their participation. First, however, we introduce heritage, performance, and penal tourism in greater depth before turning to our specific case study of the Galleries of Justice and the performative encounters of past and present that are brought to life there.

Heritage, performance, and the prison museum

Many of the buildings constructed during the nineteenth-century prison-building boom have now been rendered obsolete and unfit for use, owing to architectural degradation and the cost of maintenance. In place of their former purpose, many have been repackaged within a framework of heritage preservation, creating the recreational leisure pursuit of "penal tourism" (Strange and Kempa, 2003). Strange and Kempa illustrate the prominent examples of Alcatraz in the United States and Robben Island in South Africa that now serve as museums and heritage sites (2003); while Welch and Macuare (2011) explore the Argentine Penitentiary Museum in Buenos Aires, which operated as a prison until 1947. Walby and Piché (2011) examine how multiple penal sites such as county and local prisons as well as a former warden's house in Ontario, Canada have been utilized for tourism purposes.

Such studies have investigated penal tourism through a variety of analytic lenses. Bruggeman's (2012) examination of America's archetypal "separate system" prison – the Eastern State Penitentiary in Philadelphia – explores the politics that many creators of penal tourist sites grapple with: the obfuscation of race, power, and community that emerge via systems of preservation. Barton and Brown (2012) consider tourist travels to a functioning prison site in Dartmoor, UK during the inter-war period, where visitors arrived daily to watch inmates marching out for afternoon labor. Such work highlights the use of prison sites as "spectacles" or displays of extraordinary, "other" spaces for a fascinated public. At the Angola State Prison in Louisiana, the public pays to watch prisoners competing in rodeo activities (Adams, 2001; Schrift, 2004).

Piché and Walby (2010) problematize penal tourism, suggesting that it rarely offers an accurate insight into prison life. Such a critique hinges on the spatial and temporal distancing and distinctions that arise at such heritage sites. While such sites aim to bridge the gap between prison and society to foster greater understanding, often the reverse happens. Indeed, as the tourist engages in a boundary crossing from the "outside" to see/watch/engage with elements of the "inside," the "inside" is set up as a profoundly different or "other" space, expanding the chasm between society and prison rather than bringing the two closer together. Simultaneously, such tourist sites aim to enhance present-day understandings of carcerality. Yet, a temporal distancing also occurs. The presentation of particular penal histories is often at odds with the present shape of contemporary prisons, yet these narratives come to represent prison life itself – with common themes of punishment, servitude and rehabilitation – further extending the gap in understanding between those on the outside and inside.

In recent years, scholars of memory and heritage have explored the production and consumption of the past through processes of performance (Johnson, 1999a, 1999b) – the use of more-than-representational, embodied, and "enlivened" techniques – to bridge such gaps and bring visitors into closer "touch" with specific times and spaces in the past. As Tilden notes, "[i]nterpretation is the means through which to shorten the distance between site/artefact and the visitor" (1957: 8). Live interpretation, where actors perform to interact with visitors, ranges from "historic re-enactment through to theatre, storytelling and role play" (Leighton, 2007: 120). In recent years, this method has been especially popular at the museum (see Jackson, 2000; Malcolm-Davies, 2004; Shafernich, 1993; Wallace, 1981). As Leighton describes,

> The Galleries of Justice, the former Shire Hall and county gaol for Notting-hamshire, located in Nottingham's Lace Market, invites visitors to travel through three centuries of crime and punishment. Visitors experience a real trial in an authentic Victorian courtroom before being sentenced and "sent down" to the original cells and caves deep within the site. During their stay in the cells visitors meet the warders, costumed live interpreters who "welcome" them for their stay, along with the hangman; the ultimate experience is however the Haunted Lock-in.
>
> (Leighton, 2007: 120)

Crang, in particular, explores the contested ground that such performances elicit in museum spaces, arguing that they "create a powerful effect of realism … [that] lends authority" (2003: 8). As Schouten (1995: 21) echoes, "[v]isitors to historic sites are looking for an experience, a new reality based on the tangible remains of the past. For them, this is the very essence of the heritage experience." Indeed, performance ostensibly brings visitors closer to the histories represented, through presenting them in a non-representative manner (Thrift, 1996). Such "real-life" depictions work to fold pasts and presents together by "recreating environments"; bringing disparate times into touch through the personal, embodied, interpretive nature of the performance. This often makes the past appear more "authentic" to the visitor (who comes into closer, "genuine" contact with the past) (Crang, 2003: 266). Prison museum curators have employed these approaches to representing history. Yet for the prison museum these techniques are doubly significant in view of the subject matter. The work of the penal tourist site is not simply to narrate the past but to narrate an "other" space. Indeed, penal tourist sites increasingly use performative effects to elicit empathy with incarcerated others. For the visitor engaging with penal tourist sites like the Galleries of Justice, techniques such as embodied performance – of both staff (enacting jailers) and visitors (enacting convicts) – rely on a boundary crossing, a spatial dislocation from the world outside to a performative re-creation of the inside, traveling back and forth, from past and present through narration and character performances (see Baker, 2014; Galleries of Justice, n.d.a).

Accessing the prison museum

In 1993, at the end of its long history in penal justice, the Lace Market Heritage Trust took ownership of the site of the Nottingham Shire Hall and County Gaol. The Trust transformed the site into the Galleries of Justice museum, which opened two years later. Official records state that a jail existed from 1449, although it is believed that criminals and other miscreants were detained here from a much earlier date. In 1878, the prison was closed due to appalling conditions but the courtroom remained in use until 1986. The Galleries of Justice makes use of a variety of different spaces on its premises. In addition to the courtroom, visitors are able to explore the Georgian and Victorian prison cells; "dark" punishment cells and the oubliette (dungeon); the exercise yard; and the additional wing added in 1833. The museum itself is one which tells multiple histories, both of specific carceral pasts relating to the former court and prison on site, and a national history of crime and punishment in the UK.

Our fieldwork at the Galleries of Justice entailed eight site visits where we navigated the site and its exhibits independently, and engaged with a variety of the scheduled performance-led tours. We also participated in audio tours (available on days of the week when guided tours were unavailable) and virtual tours (designed for visitors with limited mobility, owing to the inaccessibility of the lower parts of the building). Visits were conducted over two years in order to gain a sense of how performances differed with the time of year; the weather; and, with respect to performance-led tours, the costumed interpreter leading the tour, and the dynamic of differing tour groups. This method was a form of auto-ethnography (see Crang and Cook, 2007: 6) where we shuttled between multiple insider/outside roles as researcher/tourist and civilian/prisoner in order to understand how the penal past was made usable by the museum and comprehensible to visitors. Through "assaying" the past in the present (following Garrett, 2011), we were able to reflexively and critically consider the role of performance in negotiating the complex temporalities and spatialities enfolded in the prison museum.

In addition to this, we also held conversations with museum staff, including costumed interpreters and those in a curating role; and analyzed promotional materials and guidebooks (both in print and online). This was in conjunction with collating 486 online visitor reviews of the museum. We analyzed user reviews posted during the time frame of the research (January 2012 to November 2014) in order to accumulate the most current opinions and triangulate with the auto-ethnographic observations made at the site (and therefore not referring to defunct exhibitions). These postings provided a rich and informative insight into tourist engagements with the penal museum (Langer and Beckman, 2005; Paechter, 2013).

Performing the carceral past in the present

Advertised as one of its unique attractions, the "exciting and engaging" performance-led tours are one of the museum's central methods of communicating the carceral past (Galleries of Justice, n.d.d). During these tours costumed interpreters lead visitors around different parts of the building. For example, a tour around the County Courtroom is led by either an usher or the Sheriff of Nottingham; the pre-reform cells are exhibited by a turnkey, his wife, or an executioner; and the Victorian prison and exercise yard are introduced by a prison warden or a matron. These interpreters are advertised as a substitute for the real thing, with the museum proclaiming that "costumed actors will make you feel right at home" (Galleries of Justice, n.d.d). The choice of the word "home" is particularly poignant, given that the space of the prison is set up as spatially distinct from, and "other" to, the space of the home. What is implied is that the experiences elicited on the tours are aimed to make visitors *genuinely* feel they are enfolded within the prison, at the various moments of the past that are represented and performed. As Crang notes, there is an often an "effective authority to … interpretation" (2003: 266), which the Galleries of Justice embrace to engage visitors in their narration of the past. Indeed, the job advert for a costumed interpreter at the museum stresses the need for "authenticity" and "commitment" to educational work (Galleries of Justice, n.d.c). As such, performances are intended to bridge the gap between prison and home, past and present, through the very embodied, haptic, and felt elements that performance encapsulates.

Yet the job description for staff working as costumed interpreters stresses the need to work "in line with prepared information and scripts" (Galleries of Justice, n.d.c), to ensure that performances are "genuine," "authentic," and hence hold authority as "accurate" narrations of the carceral past. This suggests that far from being a dynamic bridge between past and present, the tours simply become spoken representations of the past; rather than working as a lively and embodied means of dealing with the spatial and temporal distancing which the past prison setting presents. The scripts are peppered with well-positioned points of exclamation: the number of people executed in the building; gruesome punishments "carried out below your very feet"; and supposedly genuine artifacts from prisons across the country – all interspersed between the obligatory timeline of key dates in the history of criminal justice in the UK.

While some points were common in narration and plotline, the tours we engaged with differed radically. This depended on the make-up of each group partaking in the tour, and the set of costumed interpreters who delivered the scripts. We further found from our conversations with museum curators and costumed interpreters that scripts are often adapted by staff members themselves, contributing to a "forever changing" experience of tours (Galleries of Justice, n.d.d). As such, the narration of the temporally and spatially distinct prison past is retold and reshaped with each performance. This forever changing performance of the past is further complicated by the organization of the narrative, which is simultaneously linear but also temporally fractured. For example, a discrete, linear, historical story unfolds, related to the space in which the costumed interpreter performs his or her role. However, upon moving to another space within the museum, time reverses and the visitor is transported back again, as another linear history is told from the perspective of this new location. The narrative in the County Courtroom, for example, stretches from the late 1600s to the present day, before the tour returns to the earlier Georgian period of history once more, when this performance ends, and the visitor is routed to the prison below, and to the turnkey, his wife, or the executioner (whomever happens to be on the staff rota for that day).

This complex shuttling of time, back and forth, through the performances of the interpreters, brings strands of temporally disparate histories together, encouraging the visitor to time-travel across different eras of penal history through one interpreter whose performed character and costume, is, however, situated in only one era. Thus, while making the penal past comprehensible through direct performance, costumed interpretation also presents a palimpsest of fractured pasts that can be confusing to a visitor unable to *place* those pasts in context to penal reforms and the contemporary penal institution.

In spite of this, the majority of visitors who reviewed the museum agreed on the quality of the museum and the expertise of the costumed interpreters. Yet, adding to the complexity just mentioned, the specific content and delivery of the tours differed with the shifting position of the costumed interpreters as they oscillated between narrator and actor. Indeed, while interpreters took on specific roles couched in a particular moment in history (turnkey, executioner, matron, etc.), in telling a broader historical story within the space of their performance they shifted from this character role to that of narrator. As one visitor recounted:

> The tour was good, though as a museum professional myself I would have rather had a character either in role or just in costume as the Sheriff, not one that flitted between the two – but that's just my little bug bear and I know that everyone else in the group really enjoyed the banter.
>
> (TripAdvisor, Amanda, 2013)

Staff at the museum take on a hybrid role when they act as a guide, in the *present*, leading visitors to/from the court room, prison cells, and exercise yard; yet once in these places they switch seamlessly to a character of the *past* (a sheriff, a turnkey, a matron). Tour guides, then, embody a boundary crossing from past to present through the "emplotment of events" (to follow Crang, 1994); yet they also move from outsider to insider spatially, as they shift position from an omniscient narrator who views history from "beyond," to a character situated "within" the story.

What is most striking, however, is the power implicit in the performed narratives of the costumed interpreters (see also Bruggeman, 2012). In performing either the role of narrator or character, the costumed interpreter takes a position of power in relaying penal narratives of life on the inside. As narrator they are "overseer" of the history – placed to objectively and reliably retell a story of the past. As a character they are positioned as judges, jailers, executioners, reformers – those who can dictate control over prisoners – administering penal justice, punishment, and reform. Indeed, costumed interpreters rely on audience participation as prisoners and their previous knowledge of crime and punishment to instigate the anecdotes they tell.

Costumed interpreters themselves do not perform the role of prisoner. Indeed, the narratives told by interpreters are the stories of those whose voices have been recorded, archived, and noted as reliable sources on prison pasts. The informal histories of prisoners, those silenced through their very position, are unheard through such performances. In this sense, the history that is constantly retold by the Galleries of Justice constantly embeds the structures of dominance and subjugation that shape our understanding of the penal past. Yet, interestingly, the voices of prisoners emerge via other strategies of performance, such as through objects and mannequins that "give voice" through innovative displays (see Hoskins, 2007). In the transportation exhibition, for example, carved wooden figures are each decorated with a plaque telling "their story": giving a name, their occupation, and their role on the convict ship (Figure 5.1). More than this, quotes from "real" criminals appear throughout the museum. Conversations with staff revealed how these narratives were carefully pieced together from known historical accounts of the individuals' crimes, cross-referenced with prison journals and period-appropriate surgeons' logs, for example. As Hoskins notes, narratives of the past emerge from the vibrant and affective potentiality of an object (in this case mannequins and plaques). Rather than simply working as static items, they can speak (as well as being spoken for) and can therefore hold capacities to literally perform a past as they come into contact with visitors (Hoskins, 2007: 441). This helps to build an empathetic relation as elements of penal life emerge, bringing the visitor from the outside into touch with artifacts from the inside.

Audio recordings also play in certain exhibition rooms. These recount narratives about different elements of punishment and prison life, such as carnivalesque public punishment or the experience of nights spent in a communal women's cell. On the one hand, these recordings provide a personal narrative, but they are also often disembodied, creating a chasm or distance between the prison inside and outside. Those in positions of authority – the turnkey, matron, and so on – are given names, faces, and *character* through interpretation – Mrs.

Figure 5.1 Wooden figures portraying individuals on board the convict ship. Photograph by the authors.

Linton, Dr. Massey, and so on. The prisoner, through recordings, sometimes has no name, no character, but rather comes to represent all prisoners. This, in part, reflects a problem faced by penal museums generally, namely the few stories retained that have been told by prisoners. However, in spite of such simultaneous absence and presence (see Jones *et al.*, 2014), performances such as these facilitate an effective border crossing between two distinct spaces and times through the senses of sound and vision. One particularly gruesome audio performance appears in the "Georgian" prison, at the jail entrance. It consists of an unidentified woman burning at the stake, accompanied by the sounds of her screams, the crackling of the flames engulfing her, and prolonged choking noises. At intervals, the darkened cavern within which the mannequin is situated lights up in deep reds and flickering yellows. The brief audio states:

> She was taken from the prison barefoot and she was drawn on a hurdle to the site [background noise of shouting and jeering]. They put her on a tar barrel against the stake. She was held in place by three iron bars and a rope around her neck [screams]. The rope was pulled tight, almost strangling her [screams of pain]. They rolled the barrel away and then lit the fire [crackling of flames. Screams. Crowd shouting. Choking screams].
>
> (Galleries of Justice, n.d.a)

Furthermore, in many parts of the museum, a soundtrack of noises may be heard – objects given life through audio means. These "sounds of the prisons" are often metallic, representing irons clinking together as prisoners shuffled around. Such methods of "giving voice" to the prisoner, and prisoner experience, are particularly effective in performing the past and eliciting an understanding of the horrific elements of penal regimes at particular moments of the past. Although less dynamic and changing (compared to costumed interpretation), they work to bring the past into the present in disturbing and evocative ways that move beyond mere representation.

However, audio is not simply layered on top of objects and mannequins. In addition to these audio snippets, audio tours are substituted for the absence of costumed interpreters on Mondays and Tuesdays. On these days visitors can "roam the courts, prison and dungeons with only the voices of the past for company" (Galleries of Justice, n.d.b). Like the costumed interpreter-led tours, the audio guide shuttles back and forth from inside to outside the prison, through overarching narration and first-person character stories. Similar to the tours, the narratives shift back and forth through time, while using the sense of sound in conjunction with the architecture of the prison, objects on display, and mannequins to bring the prison to life. Where audio tours are particularly effective is in their ability to speak from a variety of first-person character stories – both jailers and prisoners. While costumed interpretation provides the visitor with a taste of discipline, punishment, and prison reform, the audio more effectively gives prisoners an active voice (see Galleries of Justice, n.d.a). Voice is particularly effective in view of its immediacy in connecting the visitor to the past penal life.

The timbres of the voice, the tone, speed, dialect, and gendering of voice provide an emotive link between the past and present, especially through the often shocking narratives relayed (see Kanngieser, 2012). For example, one audio snippet provides an insight to life in a Georgian cell. Compared to a written description, the audio, heard *in* context with the cell, enfolds past and present, inside and outside, in uncomfortable and provocative ways:

> [Croaky male voice] ... gaol fever, dysentery, low fever and the dreaded diarrhoea. We get it all down here. You get sick, there's nowhere to put you because the infirmary is being used by the bloody debtors half the time. One man died.... They didn't have anywhere to put his body so they left it out over the weekend and it stank (the) infirmary out so badly, nobody could use it.
>
> (Galleries of Justice, n.d.a)

Through such audio the museum provides both the shocking, gruesome, uncomfortable narratives of prison life, yet also those that were arguably banal, everyday, ordinary occurrences. The direct nature of performance (whether embodied, or disembodied via objects, audio, and so on), is particularly useful in attempting to bridge past and present; inside and outside. Yet, the manner of performance simultaneously provokes distance. On the audio tour, for example, voices are, on occasion, ill-fitting to the subject they portray. The actors are often well spoken, and their voices do not always seem to match the age of the character they portray. One female prisoner, 15 years of age, sounds far older (Galleries of Justice, n.d.a), compromising the sense of realism upon which the performance relies.

Visitors are also encouraged to act in ways that ostensibly mirror those of the convicted. When visiting the Galleries of Justice, entrance tickets feature randomly distributed convict numbers corresponding to a particular "real-life" criminal. Each convict number refers to boards behind closed doors that display information about the individual's crime and the date it occurred. A mirror is placed on the panel, positioning the visitor's reflection in place of the "mug shot" – transferring the visitor's visual identity to that of their convict counterpart (see Figure 5.2).

Visitors are encouraged to follow the story of "their" crime, generating a particular spectacle around the sentence their character received. Although this is an obvious attempt to shorten the distance between past and present, inside and outside via the creation of individual empathy, we found on some occasions that our convict numbers were the same as others in the group, thus removing the opportunity to truly claim them for ourselves. On various visits, "our crimes" ranged from the murder of a child to stealing a cow. On a more recent visit, one of us made the casual remark: "Oh, I'm sentenced to death again." Indeed, during our participation on these tours, the guides themselves asked members of the group questions such as "Who's due for a good whipping? Which of you is getting hanged then?" In the spirit of performing our prisoner roles appropriately,

Figure 5.2 Discovering your "real life" convict history. Photograph by the authors.

we were expected to respond to such questions, embroiling us within a created atmosphere of discipline and incarceration. After inviting us past a barred gate into the oldest part of the former prison, a man in period costume introduced himself as a "turnkey." We recorded the interchange that followed in one of our field diaries:

> The turnkey asked who has been sentenced to public whipping. I raised my hand. He smiled at me and told me it was sure to be terrible, but reminded me of the silver lining: I would only have to spend two weeks in the prison. He asked me if I had any money. I played along, I said I had none. He asked me what I thought I might be able to sell. I knew something of pre-reform English prisons. I thought to reply, "my body" but then I realised there were children in the room, maybe we don't talk about that. I shrugged my shoulders. "You've got a nice ponytail" he said, "and a full set of teeth. You can sell the hair to a wig-maker (or if it's poor quality we can use it to stuff mattresses). And the dentist loves a good supply of teeth. The rich people would rather have human than horse teeth if theirs need replacing. You can bargain before they rip them out. That should make the two weeks pass by more easily." I shuddered. Teeth. There's always something about teeth that makes me cringe. The turnkey was nonchalant. He wished me good luck for my sentence, and reminded me that my punishment was for a set number of lashes, not just until I was senseless. He hoped that I didn't pass out too soon and that I made it out alive.
>
> (Turner and Peters, 2012)

We were also expected to participate in activities relating to various areas of the prison. In many cases, penal tourist sites become just one more photo opportunity, with people lining up to enact the mundane but laborious everyday chores prisoners were assigned; to pose locked up in the pillory or stocks; or to create our own "mug shots" (see Figure 5.3). We noted some of these occasions in our ethnographic diaries:

> Today we took our own mug shots, dressing up in convict uniforms and chalked our prisoner numbers on a slate to hold up while being photographed. In this way, we were encouraged to feel some empathy with those being received to prison in the past, whilst also delighting in this experience of the extraordinary.
>
> (Turner and Peters, 2013)

For visitors with limited mobility, alternative methods of performance are also harnessed. Students at Nottingham Trent University have developed a virtual tour of the museum in conjunction with the museum staff. Using gaming technologies, visitors can "navigate themselves through virtual space" to explore the parts of the building that may otherwise only be accessed via steps or uneven ground (Museums and Heritage, 2013). The developers of the virtual model were keen to ensure that this was an interactive experience comparable to the experience provided by the live tour:

> In order to maintain a connection to the real site and engage the visitor it is imperative to also embed the site's social history within the virtual tour. For

Figure 5.3 Writing our own convict numbers before having "mug shots" taken. Photograph by the authors.

this reason it now contains audio, video and effects that provide atmosphere and narrative to tell the story of the building and give a sense of place to the 3D environment.

(Museums and Heritage, 2013)

In this experience, individuals are offered a hybrid performance. Just like their counterparts on the live tours, they perform both the role of visitor and that of prisoner experiencing these spaces. The virtual medium creates a relationship to the body via the ability of users to control their point of view in order to "move and look around the space." Similar technology is found in the first-person computer game. This "being there" is therefore vital to "incorporating a sense of presence" but it requires more than simply a lifelike reproduction of the physical environment (Museums and Heritage, 2013). As the researchers involved in the development of the virtual tour explain:

> A good storyteller can conjure up and animate the spirit of history. The power of these stories should not be underestimated – they are significant in enriching the visitor's experience of historical memories and culture and are engaging and entertaining as well as educational.
>
> (Patel and Tuck, 2008: 249–250)

In order to encapsulate these stories, the overall performance of the virtual tour includes audio, video, and other visual effects. The experience is designed to engender something close to real time, a performance of the past in the present, with flickering torches, fire, steam, videos of prisoners, and sound effects present in each of the spaces, with the visitor's movement around the 3D environment triggering the voice of the tour guide.

The politics of (co)creating understanding

Regardless of whether one experiences the museum via live interpreter, audio tour, or virtual tour, it is clear that the museum aims to foster the dual experience of entertainment and education (Galleries of Justice, n.d.d). Visitor reviews commented on the learning experience created through this kind of participation:

> "Good fun in the dock!" – Excellent museum with staff all in costume and acting as characters from the past. Visitors get roped in too and all the children in our tour party seemed to love it!
>
> (TripAdvisor, Wendy, 2013)

> I took a group of year 10 students to the Galleries of Justice and we all really enjoyed the day. The staff are fantastic and make you feel like you are really experiencing what it would have been like. The mock trial that the class trip was very well organised and the students all took part, enjoyed it and most of all learnt a lot in the process. It is definitely a must see for young and old!
>
> (TripAdvisor, NAC24, 2013)

These kinds of engagements rely on the ability of visitors to interact during their experiences. However, there is a recognition that a simultaneous "distance" must also occur in order for these visits to be successful. As Huey explains, it is the fact that although the visitor knows that pain and suffering may be occurring at these sites, it is being enacted on others (and specifically, unknown others), and in this way the sites can achieve a spectacle that is "both compelling and pleasurable" (Huey, 2011: 386). As such, the visitor can watch (and learn) about these transgressive activities, but they are not expected to experience any of the negativities (Seltzer, 1998: 271; Stephens, 2007).

However, in many cases these interactions may cross the line to affect the visitor in very physical ways. Although we, as visitors, may merely intend to look at activities happening at these sites, what we see can often make us tremble or shake, or make us feel cold or sick (Pile, 2010). As one visitor to the Galleries of Justice described:

> We then descended to the dungeon and pit areas where another female actor explained the life of prisoners in the past. It was fascinating to note that prisoners could pay for better beds and blankets, otherwise they would get thrown into the pit, which was dark and scary – I didn't dare go inside. We were then left alone to explore the area, I would say that some of the younger kids were upset and uncomfortable in that environment, and I myself wasn't quite sure where to go. Then, a "guv'nor" showed us a replica of the gallows and how hangings were performed. One of the female tourists was visibly disturbed and had to have a breather.
>
> (TripAdvisor, seantyy, 2013)

As participant tourists and pseudo-prisoners ourselves, we found elements of the performance disturbing. Being locked in a nineteenth-century cell on a cold November day, with the wind blowing through the open window bars, was chilling. So too was the sickening internal feeling evoked from the sound of a sharp crack and the chips of paint which were removed as the "turnkey" whipped his cat o' nine tails at the wall. A horrifying realization arose of the damage that would have done to human flesh. Nevertheless, as with any tourist experience, visitors to the Galleries of Justice have a choice. The room exhibiting the procedure of carrying out the sentence of death by hanging carries a warning sign, encouraging individuals of nervous disposition to bypass this particular element of the tour. The fact that visitors are "buying into" these prison experiences highlights the difference between actual prisons and penal tourist sites, and further exposes the disconnect between them which persists in spite of attempts and technologies that seek to transport the visitor to other times and places.

Conclusion

In this chapter we have drawn upon research conducted at the Galleries of Justice to explore how visitors are encouraged to "do time" or rather "do time-travel" through their engagements with carceral histories. By allowing visitors to touch, see, feel, and enact versions of penal life in the museum, individuals are able to traverse the temporal divide between then and now, and so, too, the spatial division between inside and outside. Yet, in exploring these boundary crossings, we have demonstrated how the penal museum is problematic as a heritage site – and more so, how the embodied performances of staff and visitors complicate matters further. For some visitors, enacting carceral pasts through a convict journey imprinted upon the entrance ticket – encouraging the visitor to empathize with the assigned prisoner – may promote a more attached, more manageable, more personal understanding of the prison system and its complexity. In other ways, the performance of the turnkey's likable persona, for example, provides a comic spectacle that creates a distancing effect from the harsher realities of penal history. The Galleries of Justice spectacularizes prison life as horrific, with the creation of stereotypical prison characters in the form of the costumed interpreters who provide a narrative of prison as a miserable, brutal place. However, even this horror is sensationalized to make it more agreeable to most visitors. The hearty laugh of the turnkey's wife, combined with the (largely) child-friendly narration of prisoners sentenced to execution, contributes to an atmosphere of fun, not anguish. As such, horror for entertainment prevails, and "true" horrors are sanitized.

Having said all this, the ideas of temporal dislocation are crucial here. The prison museum presents particular segments of time in penal history, choosing particular convict journeys and performances to be the focal points for visitors. These performances, and the multiple effects they evoke, produce a disconnection for the visitor – who is not simply experiencing a spatial dislocation from the world outside to a performative re-creation of life inside – but is also experiencing a temporal disconnection as they move back and forth to specific moments in carceral history. This 'time-travel' – or crossing of a temporal boundary – produces versions of penal life that render the contemporary prison as something abstract and disjointed in time. In attempting to illuminate, and make proximate and known the prison environment; the performance of particular prison journeys simultaneously creates a distance, both spatial and temporal, between the heritage site and the contemporary prison, an understanding of which they may be seeking to promote.

As such, penal tourist sites, we have proposed, play a distinctive role. This is constituted through their very nature as spaces that are not ordinarily accessible and which hold a morbid fascination. They promote a tourist experience that both seems to encapsulate "life behind bars" in the past (and present) but one that is also a (co)creation based on assumptions of that life. For the majority of liberal society, these sites of heritage and leisure rest firmly upon a support of previous perceptions of the prison, built up in media constructions and in our imaginations. This places penal tourism in an awkward conundrum, altering perceptions of a world so central to the functioning of society, yet so relatively unknown to most people. Accordingly, these are spaces that scholars of the usable carceral past must continue to explore to better understand how prison is understood and engaged with in the present.

References

Adams J (2001) The wildest show in the South: Tourism and incarceration at Angola (Louisiana State Penitentiary rodeo). *The Drama Review: A Journal of Performance Studies* 45(2): 94–108.

Baker B (2014) *Crime and Punishment: Discover Nottingham's Horrible History.* Peterborough: Jarrod Publishing.

Barton A and Brown A (2012) Dark tourism and the modern prison. *Prison Service Journal* 199: 44–49.

Bruggeman SC (2012) Reforming the carceral past: Eastern State Penitentiary and the challenge of the twenty-first-century prison museum. *Radical History Review* 113: 171–186.

Crang M (1994) Spacing times, telling times and narrating the past. *Time and Society* 3: 29–45.

Crang M (2003) On display: The poetics, politics and interpretation of exhibitions. In Blunt A, Gruffudd P, May J, Ogborn M, and Pinder D (eds) *Cultural Geographies in Practice.* London: Arnold, 255–268.

Crang M and Cook I (2007) *Doing Ethnographies.* London: Sage.

Galleries of Justice (n.d.a) *Audio Tour and On-site Audio.* Nottingham: Galleries of Justice (transcription by authors).

Galleries of Justice (n.d.b) *Crime and Punishment Leaflet.* Nottingham: Galleries of Justice.

Galleries of Justice (n.d.c) Job description: Interpreter. Galleries of Justice website. Available at: www.galleriesofjustice.org.uk/.

Galleries of Justice (n.d.d) Tours and exhibitions. Galleries of Justice website. Available at: www.galleriesofjustice.org.uk.

Garrett BL (2011) Assaying history: Creating temporal junctions through urban exploration. *Environment and Planning D: Society and Space* 29(6): 1048–1067.

Hall S (1973) *Encoding and Decoding in the Television Discourse.* Birmingham: University of Birmingham Press.

Hoskins G (2007) Materialising memory at Angel Island Immigration Station, San Francisco. *Environment and Planning A* 39: 437–455.

Huey L (2011) Crime behind the glass: Exploring the sublime in crime at The Vienna Kriminalmuseum. *Theoretical Criminology* 15(4): 381–399.

Jackson A (2000) Inter-acting with the past – The use of participatory theatre at museums and heritage sites. *Research in Drama Education* 5(2): 199–215.

Johnson NC (1999a) The spectacle of memory: Ireland's remembrance of the Great War, 1919. *Journal of Historical Geography* 25(1): 36–56.

Johnson NC (1999b) Framing the past: Time, space and the politics of heritage tourism in Ireland. *Political Geography* 18(2): 187–207.

Jones RD, Robinson J, and Turner J (2014) *The Politics of Hiding, Invisibility, and Silence: Between Absence and Presence.* London: Routledge.

Kanngieser A (2012) A sonic geography of the voice: Towards an affective politics. *Progress in Human Geography* 36(3): 336–353.

Langer R and Beckman SC (2005) Sensitive research topics: Ethnography revisited. *Qualitative Market Research: An International Journal* 8(2): 189–203.

Leighton D (2007) "Step back in time and live the legend": Experiential marketing and the heritage sector. *International Journal of Nonprofit and Voluntary Sector Marketing* 12(2): 117–125.

Malcolm-Davies J (2004) Borrowed robes: The educational value of costumed interpretation at historic sites. *International Journal of Heritage Studies* 10(3): 277–293.

Museums and Heritage (2013) Crime Punishment Virtual Tour provides access so the physically impaired. Museums and Heritage website. Available at: www.museumsandheritage.com/advisor/news/item/2801.

Paechter C (2013) Researching sensitive issues online: Implications of a hybrid insider/outsider position in a retrospective ethnographic study. *Qualitative Research* 13(1): 71–86.

Patel R and Tuck D (2008) Narrating the past: Virtual environments and narrative. In Dunn S with Keene S, Mallen G, and Bowen J (eds) *EVA London 2008: Electronic Visualisation and the Arts. Proceedings of a conference held in London 22–24 July.* London: The Chartered Institute for IT, 249–258.

Piché J and Walby K (2010) Problematizing carceral tours. *British Journal of Criminology* 50(3): 570–581.

Pile S (2010) Emotions and affect in recent human geography. *Transactions of the Institute of British Geographers* 35(1): 5–20.

Schrift M (2004) The Angola Prison rodeo: Inmate cowboys and institutional tourism. *Ethnology* 43(4): 331–344.

Schouten FFJ (1995) Heritage as historical reality. In Herbert DT (ed) *Heritage, Tourism and Society.* London: Mansell, 21–31.

Seltzer M (1998) *Serial Killers: Death and Life in America's Wound Culture.* New York: Routledge.

Shafernich SM (1993) On-site museums, open-air museums, museum villages and living history museums: Reconstructions and period rooms in the United States and the United Kingdom. *Museum Management and Curatorship* 12(1): 43–61.

Stephens J (2007) From guesthouse to Guantánamo Bay: Global tourism and the case of David Hicks. *Tourism Culture and Communication* 7(2): 149–156.

Strange C and Kempa M (2003) Shades of dark tourism: Alcatraz and Robben Island. *Annals of Tourism Research* 30(2): 386–405.

Thrift N (1996) *Spatial Formations.* London: Sage.

Tilden F (1957) *Interpreting Our Heritage.* Chapel Hill, NC: University of North Carolina Press.

TripAdvisor (2013) Galleries of Justice on-line reviews (Amanda and seantyy, November; Wendy, September; NAC24, August). Available at: www.tripadvisor.co.uk/.

Turner J (2013) *Edge and Interface: Boundary Traffic between Prison and Society.* Unpublished PhD thesis, Aberystwyth University.

Turner J and Peters K (2012, 2013) Field diaries. Galleries of Justice, Nottingham, UK.

Walby K and Piché J (2011) The polysemy of punishment memorialization: Dark tourism and Ontario's penal history museums. *Punishment and Society –International Journal of Penology* 13(4): 451–472.

Wallace M (1981) Visiting the past: History museums in the United States. *Radical History Review* 25: 63–96.

Welch M and Macuare M (2011) Penal tourism in Argentina: Bridging Foucauldian and neo-Durkheimian perspectives. *Theoretical Criminology* 15(4): 401–425.

6 Carceral retasking and the work of historical societies at decommissioned lock-ups, jails, and prisons in Ontario

Kevin Walby and Justin Piché

Introduction

Several operational jails, prisons, and penitentiaries in Canada are now 100 or more years old. Older carceral sites are sometimes decommissioned and replaced by new, larger regional facilities (Piché, 2014). Once shut down, these facilities, as well as small lock-ups[1] closed decades ago, may go unused. Despite being dormant they are difficult and expensive to pull down. On occasion these buildings are bestowed heritage status, which means they must remain standing. In many small towns and rural locations across Canada, several dozen of these decommissioned carceral sites are now surrounded by buildings, streets, commercial enterprises, and other markers of the twentieth and twenty-first centuries that could not have been anticipated when they were erected. Debates take place about what to do with such sites. Property developers sometimes propose the creation of condos, hotels, restaurants, and other residential or commercial spaces. Historical societies and heritage management groups often work hard to establish museums and other cultural sites. Indoor and outdoor tours of the facilities may ensue, often referred to as "dark tourism" because of the death and tragedy that mark incarceration and punishment (Lennon and Foley, 2000).

Little existing literature has examined how historical societies become involved in establishing museums at decommissioned penal sites, which we refer to as "carceral retasking." Carceral retasking is "the act of turning a decommissioned penitentiary, prison, jail, or lock-up into another enterprise that continues to reproduce imprisonment as a dominant idea and/or material practice" (Ferguson *et al.*, 2014: 83–84). Such retasking may involve remodeling the grounds and the architecture, repositioning artifacts in and around the site, and reframing the vision and sense of a carceral site in its host community. Drawing upon data gathered as part of a five-year qualitative study examining 45 penal tourism sites across Canada, this chapter traces new purposes assigned to carceral sites across Ontario, one jurisdiction where there is a high concentration of penal history museums (27). We focus on three sites in the province: (1) the Hillsdale Lock-up in Springwater; (2) the old Lindsay Jail or Victoria County Gaol in Lindsay; and (3) the Huron Historic Gaol in Goderich. These sites demonstrate the process of carceral retasking at earlier and later stages. These acts of restoration and reclamation add symbolic relevance to old stones and steel bars, creating local attachment to buildings that decades earlier would have been considered community eyesores or had a very different meaning in the years when they deprived individuals of liberty (see also Cheung, 2003).

Engaging with debates about heritage in historical geography and penal tourism studies, we explore the use of decommissioned carceral sites as history museums. We show that local historical and heritage societies figure prominently in these initiatives to repurpose former lock-ups, jails, and prisons through their curation work that allow tourists to have fleeting encounters with carceral architecture, space, relics, narratives, and experiences. Our main purpose is to reveal the importance of historical and heritage societies in these projects. Finally, we reflect on what our findings mean for debates about the retasking of sites of confinement and punishment.

Historical geography of carceral spaces and museum studies

Historical geographers examine the spatial organization of power historically (Livingstone, 1995; Philo, 1987), of which the carceral is a key component. In our historical geography of these sites we examine spaces, buildings, and architecture, along with changes in them over time. The spaces and buildings we investigate are all former lock-ups, jails, prisons, penitentiaries, or other carceral sites. However, analyses in historical geography may also explore issues of landscape or community, edifice or roadway, a room or section of a building (Hardy, 1988). The point is to move beyond histories of disembodied events and people to instead look at material things, places, buildings, and lands (Morin, 2013).

When conducting an historical geography of a site it is important to keep several factors in mind. As Johnson (1996) argues, the background and context must be understood, the way that narratives about the site are sequenced and arranged must be known, and the way that the site has been rearranged over time must be documented. The reactions of visitors and users in the contemporary period should also be assessed. Although we do not address the reactions of visitors in this chapter, we have in other contributions (e.g., Ferguson *et al.*, 2014). Our purpose here is to investigate what happens to decommissioned carceral sites in Ontario. In our analysis of carceral retasking we address the background of the three sites, the narratives presented in their retasked form, and the role of historical societies and heritage management groups in arranging these spaces to arrive at the shape they have taken in the present.

We use the idea of carceral retasking to refer to the process of repurposing such sites once they are shut down by state agencies. This aspect of our research explores how the remodeling, repositioning, and reframing that carceral retasking entails is done, who gets involved and why, and the outcomes. Retasking is an ongoing process, as architecture and space are preserved, restored, or modified; relics are put on display; narratives are attached to objects, people, and buildings; and new features are added to these cultural sites. Not all former carceral facilities become penal history museums. Some of these facilities become other kinds of museums and municipal buildings. Rather than memorial-izing these spaces as museums, in some cases no efforts are made to mark the past uses of carceral sites and buildings. For example, there is no mention that a home resembling a cabin located in Unionville, Ontario once served as a tempo-rary lock-up in the area starting in the mid-1850s, at a time when shipping by train brought numerous "drunks and rowdies" to what was then a small village (Brown, 2006: 106). Of course, others like the Toronto Central Prison were simply bulldozed (Brown, 2006). However, we are curious about what happens to those sites that remain standing because of heritage status or due to efforts by heritage societies and committees to save and transform them.

Carceral retasking is often carried out in the name of heritage management and preservation, and intersects with issues of nostalgia. Heritage management involves efforts to attract tourists to view museums and historical displays. As Nuryanti (1996: 250) notes, "heritage tourism offers opportunities to portray the past in the present. It provides an infinite time and space in which the past can be experienced through the prism of the endless possibilities of interpretation." Literature on heritage management indicates that a key feature of such heritage sites is the built space itself. Built heritage refers to the reclamation of buildings and monuments created centuries ago, comprising material elements that have changed little and thus provide some encounter with so-called authenticity for tourists. In turn, heritage is claimed to be fundamental to the idea of authenticity, though authenticity can take many shapes (Zhu, 2012; Martin, 2010; Taylor, 2001) and be manipulated in many ways. For instance, "original authenticity" refers to the presence of design and material incorporated in the earliest form of the site (Andriotis, 2011). Seeing the first of something is an experience that tourists seek. At the other end of the spectrum, MacCannell (1973) uses the idea of "staged authenticity" to refer to the work which tourism providers do to make tourists feel as if they are having an encounter with the authentic. The idea of staged authenticity suggests that these museums all feature "front-stage" and contrived presentations with limited, if any, genuine access to anything authentic in them.

Studies that focus on heritage suggest that nostalgia is also a key factor that drives tourism. Frow (1991) argues that tourism companies must prepare the semiotics of nostalgia. Yet, Caton and Santos (2007) challenge the idea that simple nostalgia drives tourists to visit such sites; they may be destinations that families visit on holidays just for fun, for instance (Thomas, 1989). Such visita-tions are opportunities for meaning making, as well as the imposition of new significance upon historical objects and events. Along these lines, Harrison (2001: 171) has argued that tourist experiences can give "intense personal meaning to the tourist ... not in a simple state of banal escape or vulgar con-sumption, but in a much more complex one, 'enwrapped' in intense sensual pleasure." These debates about heritage relate directly to the process of carceral retasking, insofar as tourists seek encounters with so-called authentic carceral spaces at such sites. For example, in an analysis of online comments made in response to news stories and social media posts related to tours of Kingston

Penitentiary following its closure in September 2013, Ferguson and colleagues (2014) found that many commenters wanted to gain access to the facility to experience what life was really like for Canada's most notorious prisoners.

The sites we examine below started out as operational lock-ups, jails, or prisons, but they have now been retasked as museums. Not all penal history museums are housed in former sites of confinement and punishment, but those that are tend to exploit the original carceral building and architecture left behind in their displays. As Gordon (2008) asserts, museums were once reserved for the elite, but have now become ubiquitous and take many forms. Much literature, she argues, has focused on professional museums that are curated by people with graduate degrees. However, the most common and overlooked heritage sites – often smaller in size – are put together by local people who do not have professional training in museum studies. Gordon uses the idea of "community exhibitions" to refer to museums and other cultural sites put together by local people and historical societies primarily for others in their social network and small numbers of tourists who may venture to their rural area. Reclamation of local heritage is the stated driving force behind the curation of community exhibitions. The people doing the curating are usually connected to the history they are showing and telling. Morin (2013: 10) likewise argues that small-scale and local heritage sites with "no particular capital-generating potential" should be included in any historical geography of the carceral.

From the professionally curated to the more informal and local displays, museums are not neutral arenas for the production of social meanings. For example, Bennett (1995) examines the history of professional museums and argues that by the late eighteenth and early nineteenth centuries, governing bodies were starting to consider museums as a means of social control. Exposing people to museums was intended to make them more likely to be "accessible to moral sentiments, generous feelings, and religious and devout convictions and conduct" (1995: 17–18). Penal history museums also expose people to a range of emotions and narratives concerning prisoners and punishment, which can influence how people think about prisoners past and present (Walby and Piché, 2011). These sites stray from the "white cube" aesthetic of contemporary museums meant to be as unobtrusive as possible in order to focus the visitors' attention on the displays (Giebelhausen, 2008). However, the materials left behind in the decommissioned carceral facilities (e.g., cells, walls) add to the experience. For example, prisoner writing and graffiti on the walls are treated as displays in themselves and prison cells are sought out by visitors to experience the chill of dark tourism (Wilson, 2008). Penal history sites continue to shape the sentiments and convictions of tourists, sometimes fostering punitive feelings or perhaps raising questions about imprisonment and punishment (Brown, 2009).

Carceral retasking also raises questions about the relationship between decommissioned sites of confinement and punishment, and the communities in which they are located. Hein (2000) has examined how museums situated in particular localities create diverse interpretive communities in relation to the displays. However, with many penal history museums located in former lock-ups, jails, and prisons in small towns, the museum can become fundamental to the community in more basic ways. It may be one of the only tourist attractions in town or one of the only public relics remaining from decades gone by. In this sense, heritage and reclamation have a local meaning for residents in towns not easily subsumed by overarching claims about the role of museums in contemporary society. To understand these sites and their current uses, we assess the role of penal history museums in small rural communities when the primary motivation for establishing them is reclaiming heritage.

Heritage committees and societies are uniquely positioned to engage in carceral retasking. In Ontario, these entities advise staff, conduct research on heritage sites, fundraise, and generate community interest in their historical reclamation projects. The Ontario Heritage Act (1990) established the process by which heritage societies can be created, giving them powers to do their work. These bodies therefore not only represent the community and make decisions about how to depict their past; they also engage in municipal planning, which has a material impact upon the town and region in which they operate. Thus, the work of heritage committees and societies creates a mystique for old buildings, positioning them as central players in the past of a town or region (Leary and Sholes, 2000). Historical and heritage societies likewise reclaim decommissioned carceral sites in the name of the town or region to promote a supposed shared identity and past. This does not mean that they always strive to "get it right" when it comes to the story of the local community or the story of punishment. Indeed, the focus may be on novelty or entertainment as much or more than on education and historical accuracy.

Studying Canadian penal history museums

For an initial sample, we selected penal history museums in Canada that were part of regional or provincial-territorial tourism networks, promoted on the Internet or mentioned in Brown's (2006) book on heritage "county gaols," "local lock-ups," and "big houses" in Ontario. This sample snowballed as we conducted our field visits and became aware of other penal tourism destinations. We tried to include penal history museums in large cities, smaller towns, and rural settings, including Francophone areas, to ensure diversity in our sample. Some of the 45 museums where we conducted our fieldwork were solely dedicated to narrating penal history with no other forms of operations within the sites, while others were co-located with different museum or archival spaces, hospitality (e.g., lodging, restaurants, bars), or tourism services. We have also conducted research at several smaller lock-ups. Police, "crime" or "justice" museums, or other locales such as "human rights" or military museums, were not included in our sample.

Beyond visiting these sites and documenting their architectural, spatial, visual, and narrative features, our research involved 52 semi-structured interviews with penal museum historians, curators, administrators, as well as other staff and volunteers.[2] Those interviewed were asked questions about the history of their respective museums, the original purposes of the sites, and how that information shapes curation practices. Where permission was obtained, we also observed penal history museum staff and volunteers using field notes, which were often supplemented by informal questioning. We focused on how decisions are made about the curation, as well as marketing practices that shape the meanings of penality in these museums. We also observed the narratives of staff and volunteers in their interactions with visitors during tours where offered and/or when fielding questions about aspects of penality from patrons. This approach was supplemented by archival research on newspaper stories, promotional, and other materials relevant to museum operations and history. We have also subsequently tracked special events and exhibitions that have occurred at these sites. The three sites analyzed here were selected because they demonstrate the role local historical societies play in carceral retasking.

Hillsdale Lock-up in Springwater

Heritage preservation was crucial to the carceral retasking of the Hillsdale Lock-up in Springwater, Ontario into a penal history museum. We begin our analysis with a brief overview of the history of this site before examining the process of carceral retasking.

In 1900, a local doctor requested funds from the provincial government to build a lock-up and courthouse for Hillsdale. The courthouse was not built until August 1906 when the County of Simcoe constructed a courthouse with cells along one wall. The Hillsdale Lock-up is approximately 20 by 28 feet and constructed of stone, brick, and concrete. The roof of the building is covered in pressed sheets of metal, and features a chimney-stack. The primary entrance is located on the east facade. Maintenance for the lock-up in its early days included repairing the windows and roofing, supplying the wood used in the wood-burning stove, and cleaning. In the 1920s, the building started to be used for other purposes, such as a polling station for elections. It continued to serve multiple purposes throughout the twentieth century.

In September 2000, the Township of Springwater Heritage Committee cleaned the Hillsdale Lock-up building and began brainstorming ideas about what to do with it. The building is in near original condition in spite of the sparse maintenance measures and use of the facility for other purposes. The heritage value was not lost on the tour guide interviewed: "This is the township's only historical site. All other sites are owned by private citizens ... I don't have any records, any township records, of them ever wanting it to come down." The site has been turned into a museum. As the interviewee indicated,

> We do school trips for the museum part ... we do it for Doors Open. Doors Open is a once-a-year time when the buildings that are not normally open to the public, they open them up.... We've had a couple here, they've used this place.

On September 18, 2010 the Springwater Heritage Committee organized a Heritage Day at the Hillsdale Lock-up. As the Springwater Heritage Committee communication put it, "take a step back in time to visit the Hillsdale lock-up and experience a piece of our history" (Township of Springwater, 2010). The public was invited to the lock-up to learn about its past and local history more broadly. "We're fortunate to have one in our Township," claimed the Chair of the Springwater Heritage Committee. "It's not used on a regular basis, but we'd like the public to come in and see what it's all about." During the event, members of the Springwater Heritage Committee were in attendance to answer questions and provide information to visitors on the lock-up and its history.

Inside, the walls of the main room and the cells are painted white (see Figure 6.1). There are very few displays. However, prisoner writing on the walls, some of which is scratched into the cement, may be viewed in the cells. Some of the furniture placed inside may be original. There is a plaque outside showing the designated Historic Site in the Township of Springwater. The Springwater Heritage Committee has also been looking for pictures and archival material to create a more permanent and informative display. As declared in an August 2012 communication:

> The Springwater Heritage Committee is looking for historical photographs of the Hillsdale lock-up.... If you have any historical photos of the jail or know someone who might and would be willing to share, the Heritage Committee would love to hear from you!
>
> (Springwater News, 2012)

Figure 6.1 Cells at the Hillsdale Lock-up. Photograph by Alex Luscombe.

In 2013, the roof and the windows were replaced with replicas modeled after the original design. In the spring of 2014, there was an unveiling of the renovated building. The jail was also a popular venue for a Doors Open tour.

The Hillsdale example demonstrates the process of carceral retasking at an early, formative phase. In other words, the Springwater Heritage Committee members are in the midst of arranging the displays and narratives at this decommissioned carceral site. The local historical society and their remodeling and reframing work is therefore crucial in reclaiming the building and memorializing the site in museum form. In the name of community, they are attempting to mobilize notions of heritage to reclaim a site that previously kept some people away from others in their area for brief periods of time, transforming it into a museum through which locals may become united.

Old Lindsay Jail in Lindsay

A second example of carceral retasking at a decommissioned site of imprisonment is the old Lindsay Jail, now known as the Olde Gaol Museum. While this site – like the Hillsdale Lock-up – came to take its current form due to the efforts of local residents, the Olde Gaol Museum is different in its focus on other aspects of local history and the increasing professionalization of its museum operations.

Victoria County was founded in 1861, and the old Lindsay Jail or Victoria County Gaol opened in 1863. The original small county jail could hold 20 prisoners. Many hangings took place there during the early years. Some of the executed were buried beneath the courtyard. One of the last hangings to take place within the courtyard was in 1924, when Fred McGaughey was executed following his conviction for murder. The jail continued to be used throughout the twentieth century. In the 1980s and 1990s, the jail expanded. In 1982, new wings were added to the facility, beginning with a new warden's office, as well as a new cell block. The courtyard walls were rebuilt in 1990. The jail was closed in 2003 when the Central East Correctional Centre near Lindsay – commonly referred to as one of the Province of Ontario's "superjails" (McElligott, 2008; Hannah-Moffat and Moore, 2002) – opened. At the time of closing, the old jail held 80 prisoners in a site designed for 60.

The Victoria County Historical Society was instrumental in turning the decommissioned jail into a museum. The goal of the reclamation campaign was to create more tourism traffic in the region. In 1976, the Ontario Historical Society – which sponsors provincial heritage projects and also designates when other bodies can become historical societies – allowed the Victoria County Historical Society to form. The goals of the Victoria County Historical Society are to collect, preserve, and exhibit material pertaining to the past of the region, acquire documents and records of early settlers, maintain a gallery of historical portraits and a museum, as well as publish information relative to and encourage the study of local history.

The Victoria County Historical Society raised funds and organized for several years in order to turn the old jail into a museum. Before receiving approval for the project, the Victoria County Historical Society began renovating the jail in hopes that it would someday become a museum. For instance, in 2008, the Society started to renovate the north cell block, which was erected during the 1982 jail renovation. This area was the most modern and required the least upgrading. This process was noted in a Victoria County Historical Society newsletter (2008), which stated:

> In May we began cleaning out the cells, replacing lights, scrubbing, priming and repainting the walls. Proceeds from the sale in the courtyard helped toward the materials required…. We prepared and submitted a proposal to create the first permanent display for "Victoria County" at the Museum. This was approved in the fall, and we have now received a $2,000.00 grant for a start on this display.

The old Lindsay Jail provides an example of the remodeling, repositioning, and reframing that carceral retasking entails. The Victoria County Historical Society applied for several grants to fund building repairs and upgrades, along with the creation of a digital media presentation of the collection with Internet access that would allow the museum to connect with the local community online. At their February 2008 meeting the Board of Directors formally decided that the name of the museum would be the Olde Gaol Museum in keeping with the original spelling of the Victoria County Gaol.

The jail hosted several visitors during the renovation and remodeling. On August 20, 2010, Provincial Minister of Culture and Tourism Michael Chan visited the renovation site to praise the project. On October 5, 2010, a performer named "The Dark Master" (Steve Santini) attempted an extreme escape from the old Lindsay Jail. The goal of these initiatives was to draw media attention to the renovation and to raise the profile of the Victoria County Historical Society.

During this period, the old jail was modified. The first floor underwent remodeling. The second- and third-floor areas were modified to meet fire code regulations with the installation of a sprinkler system. A new entrance to the building was created on Colborne Street West with a wheelchair ramp, steps, and modern entrance and exit doors. Changes also included the creation of an archive room where residents in the region could trace their family roots. As was noted in a Victoria County Historical Society (2011) newsletter, the hope was that "The 'Olde Gaol' will be the prime Cultural and Tourism attraction of the City of Kawartha Lakes, we are proud to call home!"

After the Victoria County Historical Society raised US$250,000 in a fundraising campaign, the closed jail was finally turned into a museum. The Olde Gaol Museum opened its doors on May 24, 2011. Several exhibitions of local history were on display throughout the building. Old prison cells were turned into exhibition rooms featuring the works and collections of famous local people. According to the newsletter (2011), "Victoria County Historical Society also had a dream.…

There were bumps in the road. But now, thanks to the dedication of many volunteers and the foresight of the City to preserve the old jail, it is a reality." The public was invited to visit the museum and to become more involved in the organization.

Not all exhibitions were ready for display when the grand opening of the Olde Gaol Museum occurred, nor when a member of our research team was there conducting interviews a few years later. Memorabilia from the First and Second World Wars, along with an exhibition of Leslie Frost's artifacts from his tenure as Ontario's premier from 1949 to 1961, are some of the prominently positioned displays. As a tour guide indicated during the interview, "Even from last year, or even from a couple weeks ago, we're constantly finding new things and putting them up and moving around displays." He continued, "The tour changes all the time because of the people that come in. We have former guards or former prisoners who know more about the history. So as they tell me stuff, I add that into the tour."

The tours also draw attention to the fact that the museum is located in a former jail where people were imprisoned, punished, and hanged. Sometimes these aspects of the site are framed in ways consistent with what has been referred to as dark tourism. Dark tourism occurs when tourists visit sites that promote representations of death and disaster (Lennon and Foley, 2000; Strange and Kempa, 2003) and that encourage tourist encounters with the macabre. For instance, the tour draws attention to the jail cells (see Figure 6.2):

Figure 6.2 Cell block at the Olde Gaol Museum. Photograph by Alex Luscombe.

INTERVIEWER: Has anybody ever come in here and had a bad experience?
RESPONDENT: Not that I can say. I don't really think so, no. Usually the kids cry when the parents want to put them in the cells, though. Other than that, not really anything.
INTERVIEWER: Do the parents often try and put the kids in the cells?
RESPONDENT: Pretty much every time. That's the first thing they always say, "Is there a cell open?"

This excerpt suggests that there are opportunities for dark tourism experiences despite the efforts of the Victoria County Historical Society to create a local history museum rather than a jail museum per se. The authentic aspects of the historic jail are reframed and staged for entertainment, as well as for ad hoc attempts at "scaring kids straight" by their tourist parents.

While members of historical societies may not be professional curators in many cases, they do use aspects of former carceral space to create other dark tourism experiences. For instance, in October 2013, the jail museum in Lindsay was used for Halloween festivities. As was noted in another Historical Society newsletter (2013):

As we transform the Museum into a haunted mansion, our upcoming Halloween Event for the Society is well underway, again we are very pleased to offer pictures behind bars for the kids at $5.00 per picture for the public's enjoyment. Treats for All.

The Halloween frights are parallel to dark tourism endeavors that memorialize death and the macabre (Lennon and Foley, 2000), but differ in that they make light of the torture and punishment that occurred at the site. The Victoria County Historical Society plays on the dark tourism aspects of the original architecture when such events are organized. It was also announced that the Victoria County Historical Society had received a two-year grant from the Trillium Foundation, which provides funds to community initiatives in Ontario. This funding allowed a museum manager to be hired.

The example of the old Lindsay Jail shows the role played by local historical societies in heritage management and reclamation, as well as some of the ways in which jail architecture is either displayed or masked to facilitate particular aims. Heritage management requires efforts at remodeling and reframing, as well as the rehabilitation of historic areas and nostalgia for long-standing communities (Nuryanti, 1996). Reframing involves representing the carceral space in a positive light to create a welcoming environment for family and community activities to take place on grounds that were once used to inflict pain upon and torment their captives. The heritage management efforts of the Victoria County Historical Society have resulted in the retasking of the old jail as a new museum for the region.

Huron Historic Gaol in Goderich

The Huron Historic Gaol located in Goderich, Ontario is an example of carceral retasking and a museum stemming from community efforts to reclaim local heritage. This site is mostly focused on penal history, and it involves significant government funding and professional staff. Below, we discuss the history of the site, and examine the long-standing efforts to transform it into a penal and general history museum for Goderich and the surrounding area.

The architect Thomas Young modeled the design of the 1841 gaol[3] using popular trends in prison reform that stemmed from the writings of Jeremy Bentham (Brown, 2006). The facility served as the area's main jail, the courthouse, and a meeting place for locals. The site was in operation for more than a century. However, with inadequate space for staff and prisoners, the provincial government struggled for nearly two decades to find a solution that would keep the jail open. The Government of Ontario closed the jail in 1972.

Almost immediately after the closing of the jail, a staff member of the museum noted that "local concern arose that it was the beginning of the end for one of the oldest and architecturally significant public buildings in Western Ontario" (Interview, July 2013). With the building facing demolition, members of the county's rural council fought to save the jail for its heritage value. A Save the Gaol Society was formed. The staff member interviewed for this study remarked that "the group of tenacious citizens attempting to save the jail faced an uphill battle, as heritage properties were not enjoying current popularity." A museum since 1974, the jail was recognized as a national historic site in 1975 and renamed the Huron Historic Gaol.

The staff member emphasized the need to focus on the building itself, rather than pack the museum with displays. She stated, "the building itself is really the artifact." Evidence of strong community ties to the site is contained in the fact that local residents donated most of the furnishings found in the museum. Community support is also apparent in the voice of one council member who was opposed to demolition:

> the most important reason is the fact that that building is the symbol for what we have evolved to be.... It's our starting point, it's the very basis of the existence of Huron County. The architecture of the building shouldn't be disrupted or changed in that regard.
>
> (Brown, 2010)

Efforts to reclaim the site thus emphasize the idea of authenticity. As recounted in one media article (Brown, 2010), the local historical society whose members include persons from the Save the Gaol Society played a pivotal role in retasking the site:

> In early 1973, about 2,000 Huron County citizens crowded into what was then called the Huron County Gaol for an open house. Former Goderich

Reeve Paul Carroll said the event was intended to get a "read on how much support there might be for retaining the building." During the previous year, a proposal was brought to county council calling for partial demolition of an exterior wall on the west side of the jail. Notably, Carroll said he was the sole voice of dissent on council, opposing the demolition.... "The amount of community support was amazing," he said. "That led to a significant change in the balance of opinion on county council when the final decisions were made" ... the community swayed the pendulum in favour of leaving the jail.... The jail was recognized as a national historic site in 1975 and renamed the Huron Historic Gaol.... Another open house was held last Monday, this time celebrating 35 years as a national historic site.

Members of Save the Gaol Society would go on to become the Huron Historic Jail Society.

Tours of the site are conducted daily throughout the summer and feature an interactive component. The curator states that rather than having a full, guided tour with someone continuously talking, the main goal of curation at the museum is creating the displays to let the "walls speak" and to let the people experience being within them. Visitors have the option of taking a self-guided walking tour, a guided walking tour, or an interactive tour accompanied by a Tour-Mate listening wand. Visitors who participate in the interactive experience take the same walking tour. At all Tour-Mate locations there are stories, "fun facts," or historical information available for access through the device. In addition to these tours, a staff member helps with two ongoing programs for youth titled *Behind the Bars* and *If these Walls Could Talk*. The former is an interactive self-guided tour involving 18 volunteers dressed up as community members who worked in the jail during the eighteenth and nineteenth centuries. Some of the displays facilitate dark tourism encounters, as noted in the following interview featured in a newspaper article, and referring to Figure 6.3:

> "It's kind of scary," added Natalie Hill, 10, referring to a prisoner exercise yard used last Monday for a rope-making activity. "When you're out here, it creeps me out to think they were out here."
>
> (Brown, 2010)

While the pursuit of authenticity is central to the operation of the Huron Historic Gaol, there is a concern among museum operators for finding other ways to generate interest in the site. For instance, the courtyard is used by members of the township for summer activities like music concerts, theater, and a variety of community programming.

The history of prisoners who were held captive at the Huron Historic Gaol offers insight into the practices of confinement and punishment of the time period. A staff member states that "the Gaol housed everything from murderers, to vagrants to debtors and even children, most importantly no distinction was

Figure 6.3 Yard at the Huron Historic Gaol. Photograph by Kevin Walby.

made between offenders; they were all seen as deviant and housed under the same roof" (Interview, July 2013). With a catalog of all prisoners kept within the Gaol one may infer what prisoners of various periods were like. Prisoners varied from Joseph Williamson, who was incarcerated for selling for auction without license, to Edward Jardine, who in 1911 was hung for murder. Other statistics available for the jail indicate that there were only four prisoners who "met the hangman." The most renowned prisoner who spent time in Huron Gaol was 14-year-old murder suspect Steven Truscott. In 1959, Truscott was arrested and tried as an adult for the murder of Lynne Harper. Truscott's case received national attention when his judgment was passed in the Huron court located directly above the cell block. A manuscript located on the wall describes the events that transpired in Truscott's case. The judge stated: "you will be kept in close confinement until Tuesday, the 8th of December 1959 and upon that day and date, you be taken to the place of execution, and that you may be hanged by the neck until you are dead." The execution never took place, as the sentence was changed to life imprisonment and Steven Truscott was moved to Kingston Penitentiary. After decades behind bars, his original conviction was later reversed when new evidence came to light (Roach, 2003). His case is now mentioned as a noteworthy example of a wrongful conviction in Canada, both at the museum and in the news media. This is one example of how such museums try to educate the public about criminalization and punishment in Canada, as well as the distinctions they make about deserving (guilty) and undeserving (innocent) prisoners.

The Huron Historic Gaol generates an operating budget in diverse ways. A significant portion of the funding comes from the Government of Canada under the community museum operating grant program. The site itself supplements a percentage of operating costs with merchandise sold at the gift shop. In addition, the site collects profits from rentals for community events such as art shows and meetings. An interviewee makes reference to contributions from the community in the form of donations, noting that donors contribute simply because "they value the heritage the site represents." She also emphasized how fortunate they are in comparison to operators of similar museums, since the county deals with most operating costs for the site. Despite having the national historical designation, there have still been budget cuts at the provincial government level. However, the site has tapped into a reservoir of graduate students seeking internships to complete their Master's studies.

Without the work of other volunteers and the community, the staff member argues, the site would not be able to stay open, revealing the importance of local involvement not only in the creation of penal history museums as recounted above, but their continued survival. Through these efforts of the Huron Historic Jail Society the jail has been reclaimed as part of the community.

Discussion

Contributing to cultural and historical geography, as well as penal tourism studies, we have highlighted three typical examples of carceral retasking in Ontario. Drawing upon these cases, we have shown one way in which local people reclaim carceral sites as a part of their heritage, as an aspect of their local community that helps them remember their formative years, along with their past family and friends. All three sites we examined involve heritage or historical societies reclaiming former carceral sites at different points in the retasking process in Ontario. In Springwater, a heritage committee only recently began to offer periodic tours and to collect items to showcase in the facility over the past few years. At the Olde Gaol Museum in Lindsay, the Victoria County Historical Society has been actively fundraising to maintain parts of the jail grounds, while transforming others for the purposes of creating a regional historical site that is now accessible to the public on a regular basis. The Huron Historic Jail Society was and continues to be instrumental in ensuring that the decommissioned jail in Goderich remains a thriving museum with different ways of encountering the walls, from self-guided tours to interactive plays.

We have argued that carceral retasking can involve remodeling the grounds and the architecture, repositioning artifacts inside the site, and reframing the prison or jail in the community. Historical and heritage societies figure prominently in these initiatives. Staff and volunteers try to allow access to the back stage of imprisonment, but the idea of staged authenticity suggests their efforts end up creating a front stage or a contrived tourist space that does not actually get tourists closer to understanding the past or present of imprisonment and punishment. For instance, at its most basic level, staff and volunteers literally sanitize the spaces to make them friendly for families and communities to reclaim and experience a few minutes in the shoes of former captors and captives.

Where the museum is concerned, Staples (1995: 45) wrote that they are "the most contrived and difficult of all heritage tourist ventures, because everything in it is certifiably dead." With penal history museums, the other layer of physical and social death is the suffering and punishment faced by prisoners. What historical societies and committees do is quite novel given these conditions. They bring these places and people back to life in a way that is sanitized and made palatable for middle-class consumers. Historical societies and committees are engaged in a "resurrectionism" and reclamation of local and regional legacies (Leary and Sholes, 2000: 66). Although, when it comes to carceral retasking, the fascinating part of reclaiming decommissioned carceral sites is how buildings that were once houses of pain and stigmatization are parlayed into venues for family fun, picnics, and other recreational tourism activities.

As we have shown, the Springwater Heritage Committee is in the process of retasking the local jail and creating a penal history museum, arranging displays there for tourists to encounter. In the case of the Victoria County Historical Society, they turned the old jail into a general history museum. With regard to the Huron Historic Gaol in Goderich, thousands of locals rallied to reclaim the jail and create a museum. The museum is now a point of pride in the community. Despite the focus on general and local history in these sites, parts of the original architecture are still used for dark tourism-type encounters, from simply allowing visitors to enter cells to more elaborate special events (e.g., on Halloween).

Examining how material carceral space is transformed through community exhibitions and local curations, we have shown how these sites become priorities for development because of the potential to draw tourists to the region. Their decommissioning and retasking is a direct consequence of the development of penality in Canada, whereby larger regional facilities are replacing these smaller jails. The "usable past" (Morin, 2013) is refashioned by local historical societies in ways that promote heritage appreciation, but which may also muddy the waters of history when it comes to comprehending the role of such facilities in Ontario's past, and explaining what imprisonment and punishment entails in the present.

Acknowledgments

Thanks to Alex Luscombe for his research work. This research was funded by the Social Sciences and Humanities Research Council of Canada (grant number 430-2012-0447).

Notes

1 A lock-up is a small one- to two-cell jail often located in a rural area where prisoners were detained overnight or for a few days before being released or transferred to a larger jail to await trial. These facilities were regularly used to imprison inebriated and quarrelsome locals to allow them to cool off (Brown, 2006).
2 In keeping with our ethical consent agreements, we protect participant confidentiality and anonymity by not using their real names or proper titles in this chapter; we simply refer to them by date of interview.
3 During the eighteenth and nineteenth centuries, "gaol" was the common spelling of jail in countries such as Australia, Britain, and Canada.

References

Andriotis K (2011) Genres of heritage authenticity: Denotations from a pilgrimage landscape. *Annals of Tourism Research* 38(4): 1613–1633.
Bennett T (1995) *The Birth of the Museum: History, Theory, Politics.* London: Routledge.
Bowman M and Pezzullo P (2009) What's so "dark" about "dark tourism"? Death, tours, and performance. *Tourist Studies* 9(3): 187–202.
Brown R (2006) *Behind Bars: Inside Ontario's Heritage Gaols.* Toronto: Natural Heritage Books.
Brown V (2010) Huron historic gaol celebrates 35 years as national historic site. *Goderich Signal Star*, 13 July.
Caton K and Santos C (2007) Heritage tourism on Route 66: Deconstructing nostalgia. *Journal of Travel Research* 45(4): 371–386.
Cheung S (2003) Remembering through space: The politics of heritage in Hong Kong. *International Journal of Heritage Studies* 9(1): 7–26.
Ferguson M, Lay E, Piché J, and Walby K (2014) Touring 'Alcatraz North': Visitor motives, reactions, and cultural representations of imprisonment at Kingston Penitentiary. *Scapegoat: Architecture, Landscape, Political Economy* 7: 83–98.
Frow J (1991) Tourism and the semiotics of nostalgia. *October* 57: 123–151.
Giebelhausen M (2008) Museum architecture: A brief history. In Macdonald S (ed) *Companion to Museum Studies.* Chichester, Sussex: Wiley, 223–244.
Gordon T (2008) Heritage, commerce, and museal display: Toward a new typology of historical exhibition in the United States. *The Public Historian* 30(4): 27–50.
Hannah-Moffat K and Moore D (2002) Correctional renewal without the frills: The politics of "get tough" punishment in Ontario. In Hermer J and Mosher J (eds) *Disorderly People: Law and the Politics of Exclusion in Ontario.* Halifax: Fernwood, 105–121.
Hardy D (1988) Historical geography and heritage studies. *Area* 20(4): 333–338.
Harrison J (2001) Thinking about tourists. *International Sociology* 16(2): 159–172.
Hein H (2000) *Museum in Transition: A Philosophical Perspective.* Washington, DC: Smithsonian Books.
Johnson N (1996) Where geography and history meet: Heritage tourism and the big house in Ireland. *Annals of the Association of American Geographers* 86(3): 551–566.
Leary T and Sholes E (2000) Authenticity of place and voice: Examples of industrial heritage preservation in the US and Europe. *The Public Historian* 22(3): 49–66.
Lennon J and Foley M (2000) *Dark Tourism: The Attraction of Death and Disaster.* London: Continuum.

Livingstone D (1995) The spaces of knowledge: Contributions towards a historical geography of science. *Environment and Planning D: Society and Space* 13(1): 5–34.

MacCannell D (1973) Staged authenticity: Arrangement of social space in tourist settings. *American Journal of Sociology* 79(3): 589–603.

Martin K (2010) Living pasts: Contested tourism authenticities, *Annals of Tourism Research* 37(2): 537–554.

McElligott G (2008) A Tory high modernism? Grand plans and visions in neoconservative Ontario. *Critical Criminology* 16(2): 123–144.

Milburn D (2010) Gaol celebrates 35 years as national historic site. *Goderich Signal Star* 7 September.

Morin K (2013) Carceral space and the usable past. *Historical Geography* 41: 1–21.

Nuryanti W (1996) Heritage and postmodern tourism. *Annals of Tourism Research* 23(2): 249–260.

Philo C (1987) "Fit localities for an asylum": The historical geography of the nineteenth century "mad business" in England. *Journal of Historical Geography* 13(4): 398–415.

Piché J (2014) A contradictory and finishing state: Explaining recent prison capacity expansion in Canada's provinces and territories. *Champ pénal/Penal Field* XI: 31.

Roach K (2003) Wrongful convictions and criminal procedure. *Brandeis Law Journal* 42: 349–369.

Springwater News (2012) August 16, p. 5. Available at: http://issuu.com/springwaternews/docs/aug_16_2012_edition_371_for_web.

Staples M (1995) Heritage tourism and local communities. *Rural Society* 5(1): 35–40.

Strange C and Kempa M (2003) Shades of dark tourism: Alcatraz and Robben Island. *Annals of Tourism Research* 30(2): 386–405.

Taylor J (2001) Authenticity and sincerity in tourism. *Annals of Tourism Research* 28(1): 7–26.

Thomas C (1989) The roles of historic sites and reasons for visiting. In Herbert D, Prentice R and Thomas C (eds) *Heritage Sites: Strategies for Marketing and Development*. Aldershot: Avebury, 62–93.

Township of Springwater (2010) Heritage Day communiqué.

Victoria County Historical Society Newsletter (2008) Vol. 3. Winter edition.

Victoria County Historical Society Newsletter (2011) Vol. 10.

Victoria County Historical Society Newsletter (2013) Fall edition. October.

Walby K and Piché J (2011) The Polysemy of punishment memorialization: Dark tourism and Ontario's penal history museums. *Punishment and Society* 13(4): 451–472.

Wilson JZ (2008) Racist and political extremist graffiti in Australian Prisons, 1970s to 1990s. *The Howard Journal of Criminal Justice* 47(1): 52–66.

Zhu Y (2012) Performing heritage: Rethinking authenticity in tourism. *Annals of Tourism Research* 39(3): 1495–1513.

7 Prisoners in Zion

Shaker sites as foundations for later communities of incarceration

Carol Medlicott

Introduction

The United Society of Believers in Christ's Second Appearing, commonly called the "Shakers," is America's oldest intentional community, dating back to the 1774 arrival of a tiny group of Shaker refugees from England. From their beginnings in America, the Shakers always sought a spatially separate sphere. The collective impulse for voluntary spatial separation has been common to many intentional communities, religious and secular alike. For the Shakers, the combined impulses of proselytization and spatial separation resulted in the establishment of over 20 communities at the movement's zenith in the mid-nineteenth century. The Shakers referred to their communities collectively as "Zion," and they maintained them as deliberately separate spheres where social, spiritual, and economic practices could be managed and sustained through distinctive spatial patterns of architecture and land use.

This chapter explores the multiple discursive and material connections between Shaker sites and spaces of incarceration. Through the development of their own compounds that combined residential housing along with spaces for economic production and religious worship, Shakers sought voluntary separation from the mainstream "world." They gradually formalized this separation and implemented it through legally binding covenants and indentures (Andrews, 1963; Stein, 1992). Along with separation came the voluntary restriction of movement: of Shakers' movements both within the village compounds and also during excursions into the world beyond. So restricted were Shakers' movements that early critics and detractors regularly accused Shaker leaders of imprisoning people, as well as of subjecting "Believers" of all ages to a range of violent abuses (De Wolfe, 1999, 2002; Sakmyster, 2011). Yet Shakers themselves earnestly resisted the notion that they were imprisoned. Given the historical and geographical evolution of Shaker village sites within a discourse of ideologically based voluntary incarceration, it is ironic that no fewer than six of the original 20-odd sites were transformed into actual sites of later incarceration at the conclusion of their occupancy by the Shakers. Moreover, several other Shaker sites were later adapted to house other intentional communities (i.e., homes for the aged, boarding schools, or religious orders).

Nicoletta (2003) contends that the built environments of Shaker settlements were carefully planned to control and regulate members' behaviors and engineer social and spiritual outcomes. I argue that because Shaker sites evolved both to actively integrate these "architectures of control" and to satisfy the multi-layered spatial demands of total institutions (Goffman, 1961), their spatial configurations were more likely to conform to the needs of other total institutions, including communities of incarceration. For Goffman, total institutions could include prisons, asylums, orphanages, sanitoriums, and homes for the aged, as well as residential religious communities. Practices of total institutions include, among others: the spatial sequestration of a large number of individuals; the concentration of the full range of social activities within that single space; the erosion of inmates' individual identities through such practices as uniform dress and use of titles; the imposition of regulated time schedules and movements on inmates; and the conviction that all practices are justified as a reinforcement of the institution itself. As applied to religious communities, total institutions stress the eradication of the self and the regimented practice of austere living (Hunt, 2002; Gardner *et al.*, 2008; Scott, 2011).

Certainly these dimensions of the total institution resonate strongly with the Shakers, who stressed conversion through personal confession of sin, ongoing public confession of even slight infractions, extreme physical exertion during worship, and renunciation of fashionable and decorative clothing, among other traits. Critics of Goffman have noted that the total institution is never truly "total" (Farrington, 1992; Moran, 2014), and that even the most sealed spaces develop functional porosity through a range of interactions between the institution's inhabitants and the world beyond the institution. Shaker sites were (and remain) similarly porous. Relations between Shaker communities and the surrounding world have been actively mediated on a range of scales – socially, spiritually, and economically. But critiques notwithstanding, Shaker communities did function, at least aspirationally, as total institutions. The manner of physical layout and structure of Shaker sites has rendered them "usable" to other total institutions. Shaker sites exhibit, quite literally, a "usable past."

Scholars from several fields are increasingly analyzing the transformation of former sites of incarceration. Many infamous prisons, asylums, and detention centers have garnered scholarly attention as they are redeveloped as heritage landscapes, entertainment or shopping venues, or residential housing (Ruetalo, 2008; Draper, 2012). Scholars are showing that the repurposing of former carceral spaces as sites of "dark" tourism can throw into sharper relief the ways in which architecture, spatial organization, social ritual, embodiment, privacy, and transparency work together to create both meaning and potency in past and present alike (Strange and Kempa, 2003; Phaswana-Mafuya and Haydam, 2005; Graham and McDowell, 2007; Flynn, 2011; McAtackney, 2013). My study draws upon a set of related but distinctive circumstances: sites associated with historically significant religious and cultural geographies arguably deserving of redevelopment attention as heritage landscapes, which are instead transformed *into* sites of incarceration. By looking at a group of people who practiced voluntary self-incarceration within spaces that were designed deliberately by the group itself to reinforce rigid and elaborate strategies of social control, we may better understand the similarities and differences between carceral communities and other collectives. Moreover, considering the practice of spiritually based voluntary self-incarceration helps to expand our understanding of the thorny relationship between mobility and power. But both approaches – examination of sites transforming from carceral spaces or into carceral spaces – are also valuable for pointing to the complexities inherent in treating embedded memories that accumulate on a site. Like former carceral sites, former Shaker sites are also palimpsests where the built landscape plays a strong role both in suggesting the past and in making the past a palpable factor that continues to be "usable" in the present.

As this chapter continues, it will first briefly address the discourse of compulsory separation that evolved within Shaker culture from the late eighteenth century. Next it will trace the manner in which the Shakers developed spaces to suit their voluntary self-incarceration and also consider how the rising conformity between Shakers and the social reform impulses of the Progressive movement contributed to the adaptive reuse of Shaker sites by other intentional communities, including carceral communities. Finally, it will survey one Shaker site that has been transformed into a site of incarceration – in Shirley, Massachusetts – to examine factors that led to the transformation. In conclusion, I will reflect on how analyzing Shaker spaces of voluntary incarceration adds a valuable dimension to carceral geographies and note some of the challenges for interpretation of these historical and spiritually significant landscapes.

Compulsory separation in Shaker discourse

Among the hundreds of hymns and songs written by the Shakers in the nineteenth century, one of the most popular began, "Farewell, farewell, vain world farewell, I find no rest in thee, Thy greatest pleasures form a hell too dark and sad for me" (Medlicott, 2011). Although at first glance it may appear to reflect the perspective of a person who is despairing or dying, in fact it reflects the normative experience of a Shaker convert. By the end of the eighteenth century, Shaker leaders based in eastern upstate New York (the movement's original locus in America) were developing the elaborate system of "gospel order" that would standardize Shaker life for the next 200 years, providing the regulatory framework for each community's spiritual and temporal status (Paterwic, 2008). Gospel order meant removing oneself to a Shaker community to live as part of a large collective unit called a "family." Within a Shaker family, biological relationships were considered irrelevant to the spiritual ties of sisterhood, brotherhood, and parenthood. Shaker brothers and sisters lived in strict obedience to a hierarchy of female and male leaders, practicing community of goods, and bound together legally by a covenant document. Part of that legal covenant included adherence to rules governing daily life, often in the most minute detail.

The nature of Shaker theology and doctrine emphasized the necessity of the Shakers' social isolation. As iconoclasts, the early Shakers fully rejected both the social and spiritual trappings of a mainstream American culture that they believed was fully in the grip of corrupt religious and political institutions. As millennialists, they believed that they were living in the "end times," that the "Christ spirit" had returned, and that their complete spatial withdrawal from the world would allow them to establish, literally, a sacred space on Earth. Shaker settlements were believed to be sinless places, inhabited not only by Shakers but also by spirits and heavenly beings (Andrews and Andrews, 1969; Promey, 1993). This contributed to the Shaker ethic of order, beauty, simplicity, and cleanliness. Because their inhabited spaces were sacred spaces, the Shakers both sought material perfection and believed material perfection was attainable (Andrews, 1966; Merton and Pearson, 2011).

Consequently, Shaker settlements existed functionally as separate small towns. US maps published in the first third of the nineteenth century are known to have used the term "Shakertown" as a generic place name for Shaker settlements in at least four different US states: Connecticut, Ohio, Kentucky, and Indiana. Within the Shaker sphere, cartographic practices reinforced the strict spatial separation between Shaker communities and the surrounding world (Medlicott, 2010). This is borne out on one early map of a then recently established Shaker community, an 1807 sketch depicting a "Section" of land purchased in southwestern Ohio by the initial group of converts in that region, together with a small contingent of eastern Shaker missionaries. The four corners of that section of land are marked by four individual crudely drawn figures that appear to be hovering angels.[1] In essence, these angel drawings reminded the map user that Shaker space was sacred space, thereby justifying and reinforcing the individual Shaker's obligation to accept voluntary incarceration within that space. Over the next 50 years, Shakers produced maps of many sites, and integrated a range of artistic techniques to emphasize their spiritual, ideological, and spatial separation from the world's people (Emlen, 1987). One ambitious Shaker even produced a formatted map collection following an extensive tour of several communities in the mid-1830s, arguably intended as the core of an atlas of the Shaker world (Medlicott, 2010).

Not only was the Shaker sphere regarded as physically and conceptually separate from the surrounding world; it was also a place to which Shakers voluntarily restricted themselves following their conversion. Ostensibly, Shakers accepted this spatial restriction willingly, a trifling price to pay in return for profound spiritual benefits. In fact, Shaker manuscripts reveal stories of converts who wished to remove to a Shaker settlement, who were distraught when family objections prevented them from doing so, and who finally completed their full spatial transition years later, when the objections were finally resolved (Patterson, 2000: 182–183). One writer expressed her permanent estrangement from her "native land" in a hymn that was widely circulated and sung in Shaker villages from east to west (Medlicott, 2011: 20–21): "My native land and shore, I'll seek no more, I'll cease to roam, Since here on Zion's ground, Where gospel fruits abound, I've found a home."

As a result of the apparent absence of freedom of movement by Shakers, one of the central accusations leveled in anti-Shaker narratives published by early detractors was that Shaker life denied individuals' basic freedom of movement. A strong theme in anti-Shaker writing from the 1780s to the 1850s consists of allegations that Shakers were under restraint, virtually imprisoned in the settlements, denied liberties by tyrannical elders and eldresses. In fact, such accusations seemed entirely legitimate for families who had lost members to the Shakers and consequently were no longer allowed to see or speak to them. Around 1810, several Shaker villages in Ohio and Kentucky were visited by unfriendly mobs, led by non-Shaker citizens from the surrounding neighborhoods who were angry at being denied access to relatives who had converted (Maclean, 1902). In such instances, the mobs accused the Shakers of kidnapping these individuals, and they demanded to see and speak with them, or in some cases, to rescue them. Possibly in response to these confrontations, pointed expressions of the spiritual freedom enjoyed by Shakers became a standard feature of the vast theological writings, social commentary, and creative literature generated by Shaker authors. Their "spiritual freedom" stood in contrast to the restrictions in daily physical movement that the Shakers regarded as trivial.

Beginning in 1821, Shaker life was carefully regulated by an elaborate set of rules known as the "Millennial Laws" (Johnson, 1967). Revised in 1845 and again in 1882, the few remaining contemporary Shakers still live under a modernized version (Andrews, 1963; Paterwic, 2008). In all of these versions, the Millennial Laws reflect extensive attention to regulating the spatial movements of Shakers, both within the settlement premises and during those occasions when journeys into the world were required. Under "Rules to be observed in going abroad & in our intercourse with the world of mankind," Shakers are told that if they "go abroad it must be with the permission of the Elders, & without such permission they must not go." Further, the Laws address the exposure of Shakers to influences in the world: "No one is allowed for the sake of curiosity, to go into the world's meeting houses, prisons, or towers, nor go on board of vessels, nor to see shows or ... to purchase or borrow books." Moreover, the Laws stipulate that any Shaker who does go abroad in the world must give a full report upon returning: "when they return home they must go and see the Elders, and give an account of their proceedings and other attending circumstances." Nor were Shakers able to receive or send uncensored mail:

> If any member ... should receive a letter from any person, he or she must show it to the Elders before it is read: also, if any member should write a letter to send abroad, it must be shown to the Elders before it is sent away.

The Laws also curtail Shakers' movements within the settlement, through a broad range of specific directives, including that individuals should not be "allowed to wander away from their companions in hand labor," that "Boys under 15 years of age are not allowed to go out hunting with guns," and that it is unwise to stop "on the road between buildings" for conversation. Clearly, with such meticulous attention to spatial "order" of a Shaker community, it is little wonder that outsiders suspected the Shakers of being virtually imprisoned.

There has been little critical scholarly engagement with the Millennial Laws of the Shakers. While many historians of the Shakers have used the Millennial Laws as a deductive tool to draw conclusions about what daily life was actually like within a prototypical Shaker community, scholars of the Shakers increasingly see the Millennial Laws as normative and prescriptive tactics adopted in response to common behaviors that went on in the communities but that Shaker leaders wished to curb. In that sense, they were like any code of regulations developed as a tool of social control. But unlike other sets of rules, the Millennial Laws had no specific penalties attached. Punishments among the Shakers mostly amounted to the withdrawal of the approbation of leaders and fellow community members. In a Society for which "Union" was the highest shared value, to "lose one's Union" and experience collective disapproval could be grave punishment indeed. Shaker manuscript journals reflect countless examples of individual Shakers earnestly working to correct the slightest infractions in personal behavior, so as to achieve perfection and avoid deeds that required confession to elders and invited collective scrutiny (Wergland, 2006; Medlicott, 2013).[2]

Inhabiting the built environments of the Shakers

Shaker communities were collective settlements whose individual maximum populations ranged from around 80 for the smallest communities, up to nearly 1,000 for the largest. All Shaker settlements began in rural settings in a wide variety of terrains, and some controlled up to 5,000 acres of land. Each community's built environment reflected the social subdivision of "family." Structurally, a Shaker family comprised a single major dwelling that housed from around 30 to over 100 people surrounded by an assortment of outbuildings and workshops where a wide range of production tasks were performed. At the height of Shakerism, some communities included as many as seven separate families.

In almost all cases, Shaker settlements grew on sites with few previous occupants, and virtually no existing structures of substantial size. Consequently, the Shakers designed and built their communities with specific intent. Several scholars have scrutinized Shaker village design, examining the ways in which the arrangement of structures contributed to the regulation of labor and creation of production zones, the management of gender interaction, the management of spirituality, and the separation between Shakers and the world's people (Emlen, 1987; Nicoletta, 1995, 2001, 2003; Kirk, 1997; Swank, 1999). The Shakers' Millennial Laws also gave attention to the built environment, stipulating building materials, exterior colors, placements of windows, and juxtaposition of structures and roadways. While rural, Shaker villages were not generally remote from major thoroughfares, and the need to trade with the world's people was constantly balanced against the need to maintain separation. Nicoletta (2003) observes that while Shakers did not adopt uniform "town planning," their villages were laid out both to shield rank-and-file members from too much expo-sure to passers-by and also to "promote trade and interaction with outsiders"; she also argues that the close proximity of workshops and residences reinforced the ability of Shaker elders to maintain surveillance and effect social control.

Shaker dwellings were designed to accommodate the highly regulated, gender-divided, celibate lifestyle of the Shakers. Within one dwelling would typically be up to a dozen or more "retiring rooms," each providing shared sleep-ing quarters for up to six or eight men or women, a large meeting room, a single dining room for collective meals, and a single large kitchen area suitable for pre-paring food for upward of 200 people. Typically, the dwelling interiors were designed to regulate interaction between genders. Twin staircases were common, as were twin doorways into dining rooms and meeting rooms, and corridors broad enough to allow people to pass one another easily without physically touching. The proportions of dwellings ranged from commodious to enormous. The "Large Brick Dwelling" of the Church Family at Union Village, Ohio was the largest brick building in the state when it was completed in 1845, using over one million bricks. Its "T-shaped" footprint matched that of many Shaker dwell-ings, including the somewhat smaller Center Family dwelling of South Union Kentucky (Figure 7.1).[3] The "Great Stone Dwelling" at Enfield, New Hampshire was the biggest single building ever constructed by the Shakers – a granite-block behemoth consisting of seven storeys and 30,000 square feet of living space.

Ideally, each Shaker family was economically self-sufficient. As the economic production systems of Shaker communities grew more sophisticated, Shaker fam-ilies developed their own distinctive types of finished goods produced (Miller, 2007). The built environment was designed to maintain each family's daily domestic needs, while also facilitating collective economic production. A near ubiquitous feature of each Shaker family was a large laundry building, usually fitted out with boilers, mechanical washing machines, drying racks, and ironing tables. Food preparation and storage required a variety of outbuildings. Separate

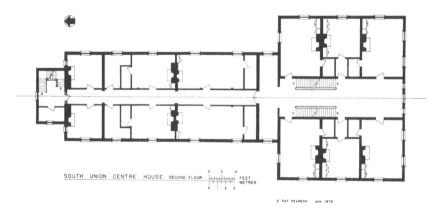

SOUTH UNION CENTRE HOUSE SECOND FLOOR

E. RAY PEARSON   JAN. 1970

Figure 7.1 Center family dwelling of South Union, Kentucky, 1970. Photograph courtesy of the Historic American Buildings Survey.

infirmaries were usually maintained, often adjacent to herb gardens to support the preparation of herbal medicines. Textile preparation played a major role in both domestic and economic production, and it required separate workshops for weaving, spinning, dyeing, and tailoring. Most Shaker villages had extensive facilities for woodworking, as furniture making was a major operation – both for commercial sales and for the community's use. The range and extent of agriculture meant that barns of all kinds were not only present in a Shaker settlement, but were quite large, efficiently designed, and well appointed. Indeed, many of the Shakers' cattle barns have achieved considerable renown among scholars of American farm architecture (Stiles, 2013).

Shaker settlements also reflected accommodation to social and spiritual needs. Each settlement had a large central meeting house, where worship was held on Sundays. In large villages, individual families took turns to use the meeting house at appointed times, while in some villages the families worshiped together. The main floor of the meeting house sometimes included perimeter seats for visitors, but always consisted of a single cavernous room, with no central posts or pillars. This design allowed unimpeded dancing during worship. Children made up a significant part of most village populations. Although celibate, many Shaker converts brought in families with them, including infants and children of all ages. And despite the practice of celibacy, a surprising number of members were "born Shaker," because early converts commonly included pregnant women. As the nineteenth century continued, orphans and child indentures were increasingly common, and in many areas Shaker villages performed an important social service by taking in unwanted children. Children were generally integrated into the village through separate "orders" – dwellings where children under the age of around 16 were housed under the supervision of an appointed group of male and female adult caretakers. It was in the "children's order" or "school order" that the vocational training of children began (Andrews and Andrews, 1973; Wergland, 2006). Almost all villages had a schoolhouse, and girls and boys attended school at separate times.

By the late nineteenth century, it was clear that some of the evolving American ideas about social rehabilitation and social perfectibility were beginning to parallel views that had long been held by the Shakers. Many groups in nineteenth-century America were developing architectural landscapes to support regulated systems of collective living, as well as social reform initiatives (Hayden, 1976; Rothman, 1971; Skotnicki, 2000). Scholars have pointed out that the ideas of early social theorists who later influenced the Progressive movement shared astonishing convergence with Shaker ideas about the best ways to curb negative human impulses and achieve social control, including manipulation of the built environment (Brewer, 1986). Indeed, some scholars argue that Shakerism strongly anticipated American social Progressivism by more than a century, with its emphasis on moral human progress being an achievable goal through the imposition of a deeply structured material and social order (Whitson, 1983; Kolmer, 1998). Scholars of Shaker architecture and design have observed the many similarities between the Shakers' built environments and those of social institutions supported by Quakers and other religious groups, noting that the expansion of Shakerism and the physical growth of Shakers' built environments "paralleled the rise of institutions of reform and confinement in the United States of the same period" (Nicoletta, 2003).

Like many elements in the mainstream Progressive communities, Shakers engineered their built environment prescriptively, to achieve behavioral and moral outcomes. In this sense, they unwittingly overlapped the nascent movement in America for institutional reform of prisons and asylums. In fact, many Shaker dwellings bore uncanny architectural similarities to contemporary edifices that were then emerging as center-pieces of leading social institutions designed for incarceration of the sick or insane. For instance, the main Shaker dwelling or "Main Building" at Enfield, Connecticut (Figure 7.2) strongly resembled the New Hampshire Asylum for the Insane (Figure 7.3) built in the same period.

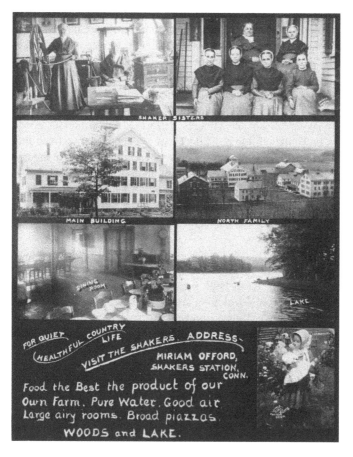

Figure 7.2 Main Shaker building, Enfield, Connecticut. Postcard courtesy of Hamilton College Library Special Collections.

NEW-HAMPSHIRE ASYLUM FOR THE INSANE.

Figure 7.3 New Hampshire Asylum for the Insane, 1843. Image courtesy of *New Hampshire Magazine 1* (November 1843: 73).

Like the social reformers working for radical new treatments of criminals, the sick, or the insane by removing them into a separate sphere where work and rehabilitation could converge, the Shakers believed that spiritual rehabilitation could be achieved in the separate sphere of a Shaker village, where worship, work, and recreation could unfold in a carefully designed space. Large Shaker communities were considered important tourist destinations from the mid-nineteenth century. For many Americans, a visit to the self-incarcerated Shakers was a voyeuristic experience offering tantalizing glimpses into an otherwise hidden world (Wergland, 2007, 2010). Shaker sites figured in the itineraries of illustrious foreign visitors such as Charles Dickens and Alexis de Tocqueville, who wished to visit "groups trying to create a better model for society" (Nicoletta, 2003: 354) As Nicoletta notes, Shaker sites were "on par with the new prisons and asylums being constructed in America."

Toward progressive-era transformation of Shaker sites

Voluntary incarceration notwithstanding, Shakers were not immune to the social changes sweeping America in the latter half of the nineteenth century. Maintaining economic viability in a rapidly urbanizing and industrializing America forced the Shakers to develop robust marketing strategies, all of which expanded and complicated the connections between Shakers and the world beyond their sphere. In short, late nineteenth-century Shakers were not as spiritually or socially sequestered as their earlier predecessors had been. By the turn of the twentieth century, Shakers were not only aware of the ongoing Progressive-era social and cultural trends in America; many were actively participating in and contributing to these trends. For instance, many Shakers had become quite active in the Peace Movement, and many were outspoken on issues of environmental conservation and the rights of women and children. A Shaker-operated periodical, *The Manifesto*, showcased these and other issues during its nearly 30-year run between 1871 and 1899. Other Shaker publications were distributed nationally, such as the annual *Shaker Almanac*, which dispensed advice on agricultural, social, and spiritual topics, and numerous Shakers published work in mainstream magazines and newspapers, which included commentary on a range of social and cultural issues.

The Shaker demographic profile by the late nineteenth century was also a factor that propelled certain kinds of transformations of Shaker sites. As the conversion of large family groups subsided, Shakers focused increasingly on cultivating the orphans and child indentures in their midst as potential new members. These children were given the option of formally becoming Shakers upon reaching legal adult age, and many did choose to remain, but at most locations, the attrition rate of males was higher than that of females. As young adults matured in Shaker communities, more males tended to be drawn to leave for marriage or for job opportunities. As females outnumbered males, more and more hired help was needed to sustain the expansive farming operations and other physical labor required at most sites. Male children and youth who "grew up Shaker" interacted far more with hired help simply because of the traditional gendered nature of work in Shaker villages. Consequently, male children had fewer opportunities to develop bonds with Shaker male mentors and role models than did the female children. This reinforced the tendency of younger adult males to leave and contributed to a general "feminization" of Shaker society. By the 1890s many Shaker communities were composed almost entirely of female members, with hired men performing much of the necessary farm work (Brewer, 1984). Many of these Shaker women grew quite involved in the Progressive-era reform-oriented social causes that appealed to American women more broadly: education, welfare of children, welfare of the sick and indigent, and rehabilitation of "wayward" people. Not surprisingly, the woman-led and woman-influenced Shaker communities were open to the adaptive reuse of their sites for purposes that promised social benefits.

At the same time, late nineteenth-century Shaker settlements sought new ways of capitalizing on their unique amenities, to attract potential converts and generate revenue. While Shaker settlements achieved prosperity and brought large tracts of land under increasingly complex forms of use during the nineteenth century, America was becoming generally more urban and industrial. Rural and agricultural spaces were drawing city dwellers as health resorts and vacation spots. Fresh air and wholesome outdoor activities were increasingly seen as antidotes to a wide range of social ills. Many Shaker villages began welcoming short-term guests who wished to enjoy a recuperative holiday. The location of several villages in parts of rural New England – such as the Berkshire mountains of western Massachusetts and eastern upstate New York – where summer tourism was growing in popularity anyway made this a reasonable strategy (Sasson, 2007). Sites outside of New England also capitalized on this notion. Most Shaker villages were located within a day's journey by carriage, car, or train from a medium or large population center, and these became benefi-ciaries of the message to urban dwellers to leave the city for restorative stays in the bucolic countryside. From Cincinnati, Lexington, and Dayton in the Ohio Valley region, to Rochester, Albany, Concord, Hartford, Boston, and Portland in the northeast, urban dwellers in many areas had convenient opportunities for exposure to the amenities of a Shaker site. The large, expansive grounds of a Shaker settlement were instantly recognizable to the visitor, due to the attractive condition of the fields, fences, and structures, as one visitor reported about his arrival at an Ohio Shaker site in 1903 (Maclean, 1907: 227–228):

> Although I had never seen the Shaker lands, the moment I struck them, I knew I was on their possessions. The fences were in good condition, the lands cared for, and there was the general aspect of thriftiness.... The view is pleasing to the eye and furnishes ample material for a beautiful landscape painting.

Along with myriad early tourism enterprises in late nineteenth-century America, Shaker villages advertised their attractive grounds by marketing postcards and enlisting photographers to produce stereoscopic views (Hoelscher, 1998; Star-buck and Swank, 1998). Many of the Shaker postcards integrated images of Shakers at work and leisure; exterior and interior views of Shaker buildings; and scenic views of farms, lakes, and woodlands. The caption of a postcard view of the Enfield, Connecticut Shaker community read, "For quiet healthful country life visit the Shakers.... Food the Best the product of our own Farm. Pure Water. Good air. Large airy rooms. Broad piazzas. Woods and Lake" (Figure 7.2).

So it is little surprise, then, against a cultural backdrop of Progressive-era strategies for effecting social change together with the rising awareness of the amenities in rural places, that Shaker sites would appeal to planners of other intentional communities and collective residential enterprises. Shaker sites offered extensive built infrastructure for housing, agricultural production, and small manufacturing, and Shaker sites were notoriously well maintained. As Shaker populations dwindled and village leaders struggled with their options, giving consideration to proposals for redevelopment as other kinds of collective institutions – both private and state funded – was both obvious and strongly alluring.

Considering post-Shaker transformations to carceral spaces

A century after their initial expansion phase, Shaker populations were in serious decline nearly everywhere. Individual communities began to consolidate their multiple families. Eventually, some villages began to close, with remaining Shakers relocating to the larger remaining villages. In the process, Shaker leaders tackled decisions about the disposition of vacated lands and buildings to new owners. In the late nineteenth century, authority in the Shaker world remained hierarchical. The ministry at the parent community of Mt. Lebanon, New York held significant influence, if not outright authority, over decisions to vacate and sell Shaker properties. Because legal ownership of lands and buildings was held formally by the appointed "trustees" of the Shaker society, the sale of vacated sites could be managed from a distance by Shaker leaders at the parent community. Finding buyers for sites designed to house collective communities of several hundred people was difficult. In one case, a Shaker site was purchased by a land developer expressly for redevelopment as a suburban residential area, with all the structures razed and the landscape extensively modified (Mollyneaux, 1987). In a few cases in very small communities, the sites were subdivided, the land and structures sold off piecemeal, and smaller structures adapted to private family homes (National Park Service, 2013).

The most common outcome was the adaptive reuse of Shaker sites for a variety of other intentional groups that intended to continue the sites' functions along the lines of a "total institution." Several of these new users were religious groups. Catholic orders took over all or part of four Shaker sites. One was taken over by a major Protestant denomination and redeveloped to include both an orphanage and a retirement home for aging missionaries and clergy, and in the mid-twentieth century, one was taken over by a communal group of Sufi Muslims. But at six sites in particular – Lower Canaan, New York; Groveland, New York; Shirley, Massachusetts; Enfield, Connecticut; Watervliet, Ohio; and Union Village, Ohio – the post-Shaker use of the sites involved their adaptation, in whole or in part, to compulsory confinement of some sort. Of those six, four are now maximum-security prisons, one is a juvenile detention center, and one has been entirely redeveloped as a corporate industrial park. Most Shaker site transformations unfolded during a protracted period between the mid-1880s and the 1970s, with the majority occurring between the early 1890s and 1920, a time of great contraction across the Shaker world.

A brief summary of circumstances at the six Shaker sites that came to be associated with communities of incarceration provides a useful context for these adaptations. At Shirley, Massachusetts; Enfield, Connecticut; and Union Village, Ohio, land from the Shaker sites was ultimately purchased by the states for the establishment of prison farms, which would both house and employ prisoners, as well as provide foodstuffs for a broader prison system. All three of these sites were later modified and hardened into higher security facilities, with the agricultural land uses scaled back or eliminated. At Watervliet, Ohio, the land and buildings were divided. Part of the land was sold to a Catholic order for the establishment of a residential community and part was sold to the nearby Dayton State Hospital, which had been founded in 1855 as a "lunatic asylum" (Frame, 1977). The addition of a working farm allowed inmates to engage in occupational therapy as well as growing some of the hospital's food. At Groveland,

New York, the Shaker land and buildings were purchased by the state for the establishment of a state asylum for epileptics, along with farmland for vocational employment of the inmates (Trompeter, 2012). That asylum was eventually closed in the 1980s and the site taken over by the New York State Department of Corrections and redeveloped as a maximum-security prison. Finally, Lower Canaan, New York was purchased privately with funds from a philanthropic couple, initially as an "industrial school for wayward boys." "Burnham Industrial Farm" evolved into a leading residential child welfare agency, Berkshire Farm Center and Services for Youth, which it remains today.

No overriding motive led any Shaker authorities to effect the transformation of multiple sites into carceral spaces. At the six different sites that became carceral spaces, individual circumstances were quite unique but certain themes across them are nevertheless common. By the 1890s, Shakers everywhere realized the inevitability of their movement's numerical decline. But aging Shakers throughout the many individual sites, east and west, also shared a strong conviction that the best way to leave a legacy in their respective regions was for the sites to be adapted for some future "useful purpose." And in the climate of Progressive-era American culture, a useful purpose meant one whereby social change could be effected through rehabilitation and education or spiritual nurturing.

Shaker records suggest that the prospect of vacating their settlements to make way for another religious community was considered a satisfactory option. Use by an active celibate order, such as a Catholic order, was appealing, as it promised the continued use of the space by a community with a strong ethic of orderly personal actions and daily devotional practice along lines broadly consistent with Shaker values. Use for the education of children, rehabilitation of "wayward" children, and care of the sick and elderly were all entirely satisfactory to the Shakers, because, in a practical sense, these functions were similar to the late nineteenth-century orientation of most Shaker communities. In several cases, Shakers continued to reside at sites for several years as the post-Shaker transformation got underway, little disturbed by the new patterns.

Communities of incarceration were not entirely objectionable to the Shakers. For the states looking for spaces to establish "prison farms," reform schools, or colonies to house the sick, Shaker sites were an obvious and attractive alternative. There were vast tracts of land already under cultivation. There were numerous farm buildings, storage buildings, workshops, and mechanical infrastructure such as mills and waterworks and woodworking machinery. There were numerous residential buildings, already configured to allow orderly housing and feeding of multiple adults. Most Shaker sites were well situated from the point of view of state authorities charged with designing correctional institutions during that period. They were accessible to main roads and large population centers, but not so close as to expose prisoners to urban dwellers. The grounds and farms provided ideal settings for compulsory labor by inmates. And during the Shakers' long habitation of the sites, most had developed in such a way that the boundaries could be clearly marked, maintained, and fortified.

For the idealistic Shakers of that period, seeing their homes redeveloped as carceral spaces was not inconsistent with their values. The notion of the potential perfectibility of incarcerated men and women, given the right circumstances, was just as realistic as the improvement of children or of the sick. The Shakers' underlying conviction of the sinful nature of all humanity, and each person's potential for redemption, contributed to a sense of optimism that the sites adapted to prison farms could indeed serve society in an ultimately positive way. The Shakers who negotiated the transitions of the sites destined to become carceral spaces were assured that these lands and grounds were "ideal" for the future purposes.

Shirley, Massachusetts: from reform school for boys to maximum-security prison

The Shaker village established at Shirley, Massachusetts in 1790 was among the earliest Shaker sites in America, but it was also destined to remain one of the smallest, with an average population that hovered around 75, except for a brief peak of about 120 in 1850. Its location is 50 miles west of Boston on the far western edge of Middlesex County. By 1900 only 12 Shakers remained at Shirley, all elderly women (Paterwic, 2008). In 1908, the remaining Shakers moved to other Shaker communities, and the Shirley site – including all buildings and nearly 800 acres of land – were sold to the State of Massachusetts for conversion to an "Industrial School for Boys" (MCRIS, 1996). This institution opened in 1909 with 100 inmate boys residing in Shaker buildings. The various Shaker workshops, barns, and outbuildings were all immediately put to use by the school (Figure 7.4).

The resident boys received intensive instruction in a range of agricultural and vocational skills, including carpentry, masonry, and orchard care. By the school's second year, plans were underway to expand the facilities to accommodate more boys by remodeling some of the original Shaker buildings and constructing new ones. The school's spokespersons claimed that its success derived from the wholesomeness of the site's natural and architectural features (Massachusetts Industrial School for Boys at Shirley, 1911):

> The change in many of these boys has been very marked, due in part to wholesome food, fresh air, regular hours of rest, and in part realizing that a fair day's work is expected of them.

The Shirley Shakers would probably expect that residence at their former home would produce a positive outcome for these inmate boys. In fact, the stated reasons for the "temporary segregation" of the Industrial School's inmates – namely to provide "curative and deterrent treatment, by training them in their minds, in their muscles, and in their morals, so that they may be returned to the community sufficiently trained" – were uncannily similar to the Shakers' longstanding approach to orphan care. But as the Industrial School increased in size

Figure 7.4 Cottages nos. 1 and 3 and Cook House, Industrial School for Boys, Shirley, Massachusetts, 1911. Postcard courtesy of Hamilton College Special Collections.

to far outpace the size of the maximum Shaker population that had ever occupied the site, it soon outgrew the existing Shaker structures. The Shaker buildings were not large or modern enough to serve the institution's needs, and the site was augmented with new structures for residences, workshops, and recreation.

By the late 1960s, the Shirley Industrial School was part of a network of ten carceral institutions operated by the Massachusetts Department of Youth Services, and it was the second largest, with 160 inmates aged 15 to 18. A series of scandals at various institutions within the network led to a reassessment of the role of incarceration in juvenile rehabilitation by Dr. Jerome Miller, a director newly appointed to head the Department of Youth Services (Mendel, 2013). In the interests of a "thorough housecleaning," Miller implemented a program of aggressive reforms that sought to phase out the more hierarchical and authoritarian "custodial model" of inmate supervision, considered outdated according to changing standards in the social work profession (Behn, 1976). But because of entrenched staff resistance to the reforms, Miller abandoned the reform efforts altogether and instead ordered the abrupt closure of all the institutions, including the Shirley Industrial School for Boys, in 1973. Still owned by the State of Massachusetts, the site was placed under the authority of the Massachusetts Department of Corrections for use as a "pre-release" residential facility for adult male prison inmates who were headed for full release (Martin, 1976). In that capacity, the historic Shaker structures provided a "halfway house" of sorts for dozens of former inmates from other prisons. At the same time, a successful grassroots effort to secure national recognition for the historical significance of the site's remaining Shaker structures resulted in the site being placed on the National Registry of Historic Places.

In 1990, due to an increasing prison population, the Massachusetts Department of Corrections curtailed the pre-release use of the Shirley site and built new minimum- and medium-security prison facilities there. Most of the historic Shaker structures remained standing at that time, though some had been moved (MCRIS, 1996). More new construction occurred in 1996 with the addition of the maximum-security Souza-Baranowski Correction Center. At the same time, the National Registry status of the Shirley Shaker site was expanded to include historic recognition for the site's use by the State of Massachusetts to achieve progressive social welfare goals as the earlier Industrial School for Boys (MCRIS, 1996). The nomination argues "that the presence of the Maximum Security and Medium Security Prison facilities, though obviously considerable, do not detract from the overall integrity of either the original Shaker village, nor of the 'industrial school campus' that overlaid it." The Shirley Shaker Historic District remains a recognized element within the "Shaker Historic Trail" marked by the National Park Service (National Park Service, 2013). Visits to the site are highly restricted however, and coordinated through the local historical society in the adjacent village of Shirley.

Conclusion

Shaker sites have proven highly adaptable as sites of incarceration. In all cases – whether religious intentional communities or carceral communities – the same pragmatic reasons applied. The sites were engineered by the Shakers to suit the requirements of a community that needed to house and feed a multitude of people in settings of order and surveillance-based accountability, where expansive land and facilities could offer comprehensive on-site employment, socialization, and recreation. For the Shakers, the spiritual requirements of their lifestyle involved voluntary self-incarceration within these highly engineered sites. In short, the Shaker communities of the past conform to the more recent notion of total institutions. It is precisely because Shakers designed their built environments to support voluntary incarceration, along with highly regulated lifestyles characteristic of total institutions, that Shaker sites have proven so well suited for transformation to other total institutions, such as carceral spaces.

Interpretation of the Shaker legacy is challenging today, because it is a piecemeal enterprise undertaken at multiple sites in many different states. Each existing Shaker museum tells only a segment of the Shaker story, and such a fractured and multifaceted approach renders a "big picture" understanding of the meaning and implications of Shaker experience in America virtually impossible. The Shaker legacy most assuredly needs to be considered on the big picture scale. Shakers comprised a vast 1,000-mile-wide intentional community that was subdivided into more than 20 separate intentional communities – separated spatially, yet highly integrated culturally and spiritually. From the many emerged the one collective. In this sense, Shakers mirror the spatial paradox of nation of which they have always been a part.

The story of the redevelopment of Shaker sites into other intentional collectives, to include carceral sites, needs far greater attention. This aspect of Shakerism's late nineteenth- and early twentieth-century phase should be integrated more fully into the public's interpretive experience at all Shaker sites. Appreciating the ease with which Shaker sites were transformed to accommodate other total institutions, including carceral communities, both underscores the distinctive features of Shaker use of space and reinforces the manner in which carceral communities bear similarities to other collectives. Shakers' innovative contributions to American material culture have recently been celebrated anew through a series of books and museum exhibitions (Miller, 2010). Many people are developing a renewed appreciation of just how many "usable" innovations may be attributed to the Shakers. The adaptive reuse of sites from the needs of a self-incarcerated people to those of a formally institutionalized or incarcerated collective demonstrates that the "usable past" can be present in the most literal sense.

Notes

1 This distinctive hand-drawn 1807 map is held in the collection of the Shaker Museum and Library, Old Chatham and New Lebanon, New York. It is pictured in Medlicott (2010: 137) and Emlen (1987: 35).
2 Wergland discusses the "perfection and overwork" of Shaker brethren who labored tirelessly to fulfill the expectations of the meticulous and pious Church Family at one Shaker village (2006: 143–158). Medlicott presents the brooding self-criticism of one Shaker sister, who was ultimately promoted to the top female leadership position in the Shaker world (2013: 193–198).
3 Many Shaker dwellings consisted of a center-hall front wing set broadside to the street or road with a long ell (perpendicular wing) extending from the rear, giving the overall structure a T-shaped footprint. This layout appeared to have begun in the western communities of Ohio, Kentucky, and Indiana, though it also spread to several Shaker villages in the eastern states. The rear ell portions may have begun as separate kitchen wings that were later attached to the main structure, with the addition of sleeping rooms on the upper storeys of the ells. In at least two cases at Union Village, rear ells not only contained kitchen facilities, but also enclosed earlier wells, making water available in the kitchens. As Shaker settlements generally sat alongside major roadways, the T-shaped dwelling offered the convenience of placing the majority of sleeping rooms in the rear portion of the structure, providing Shakers with an added layer of insulation from worldly influences (see Nicoletta, 1995).

References

Andrews ED (1963) *The People Called Shakers: A Search for the Perfect Society*. New York: Dover Publications.
Andrews ED (1966) *Religion in Wood: A Book of Shaker Furniture*. Bloomington, IN: Indiana University Press.
Andrews ED (1971) *The Community Industries of the Shakers*. New York: Emporium Publications.

Andrews ED and Andrews F (1969) *Visions of the Heavenly Sphere: A Study in Shaker Religious Art.* Charlottesville: University Press of Virginia.

Andrews ED and Andrews F (1973) The Shaker Children's Order. *Winterthur Portfolio* 8: 201–214.

Behn, R (1976) Closing the Massachusetts Public Training Schools. *Policy Sciences* 7(2): 151–161.

Brewer PJ (1984) The demographic features of the Shaker decline, 1780–1900. *The Journal of Interdisciplinary History* 15(1): 31–52.

Brewer PJ (1986) *Shaker Communities, Shaker Lives.* Hanover: University Press of New England.

De Wolfe E (1999) A very deep design at the bottom: The Shaker threat, 1780–1860. In Schultz N (ed) *Fear Itself: Enemies Real and Imagined in American Culture.* West Lafayette: Purdue University Press, 105–118.

De Wolfe E (2002) *Shaking the Faith: Women, Family, and Mary Marshall Dyer's Anti-Shaker Campaign*, 1815–1867. New York: Palgrave Macmillan.

Draper S (2012) *Afterlives of Confinement.* Pittsburgh: University of Pittsburgh Press.

Emlen RP (1987) *Shaker Village Views.* Hanover: University Press of New England.

Farrington K (1992) The modern prison as a total institution? Public perception versus objective reality. *Crime and Delinquency* 38(6): 6–26.

Flynn MK (2011) Decision-making and contested heritage in Northern Ireland: The former Maze prison/Long Kesh. *Irish Political Studies* 26(3): 383–401.

Frame R (1977) *Craig MacIntosh's Dayton Sketchbook.* Dayton, OH: Landfall Press.

Gardner P, Williams J, and Sadri M (2008) Peoples Temple: From social movement to total institution. In Dentice D and Williams J (eds) *Social Movements: Contemporary Perspectives.* Cambridge: Cambridge Scholars Publishing, 19–27.

Goffman E (1961) *Asylums: Essays on the Social Situations of Mental Patients and Other Inmates.* New York: Anchor Press.

Graham B and McDowell S (2007) Meaning in the maze: The heritage of Long Kesh. *Cultural Geographies* 14(3): 343–368.

Hayden D (1976) *Seven American Utopias: The Architecture of Communitarian Socialism, 1790–1975.* Cambridge, MA: MIT Press.

Hoelscher S (1998) The photographic construction of tourist space in Victorian America. *The Geographical Review* 88(4): 548–570.

Hunt S (2002) Between Zion and Babylon: The application of the total institution model to a Christian charismatic community: The case of the Jesus Fellowship. *Communal Societies* 22: 99–111.

Hurd HM (1916) *The Institutional Care of the Insane in the United States and Canada, Vol. 3.* Baltimore, MD: Johns Hopkins University Press.

Johnson T (1967) The Millennial Laws of 1821. *The Shaker Quarterly* 7: 35–58.

Kirk JT (1997) *The Shaker World: Life, Art, Belief.* New York: Harry N. Abrams.

Kolmer E (1998) A New Heaven and a New Earth: The progressive Shakers and social reform. *Hungarian Journal of English and American Studies* 4(1/2): 253–275.

Maclean JP (1902) Mobbing the Shakers of Union Village. *Ohio History* 11(1): 108–133.

Maclean JP (1907) *Shakers of Ohio: Fugitive Papers Concerning the Shakers of Ohio, With Unpublished Manuscripts.* Columbus, OH: FJ Heer Printing Company.

Massachusetts Industrial School for Boys at Shirley (1911) *Second Annual Report of the Trustees of the Industrial School for Boys at Shirley.* Boston, MA: Wright & Potter Printing.

McAtackney L (2013) Dealing with difficult pasts: The dark heritage of political prisons in transitional Northern Ireland and South Africa. *Prison Service Journal* 210: 17–23.

MCRIS Shirley Shaker Village (1990) *National Register Criteria Statement*. Springfield, MA: Massachusetts Cultural Resource Information System.

Medlicott C (2010) "We live at a great distance from the Church": Cartographic strategies of the Shakers, 1805–1835. *American Communal Studies Quarterly* 4(3): 123–160.

Medlicott C (2011) *"Partake a Little Morsel": Popular Shaker Hymns of the Nineteenth Century*. Clinton, NY: Richard W. Couper Press.

Medlicott C (2013) *Issachar Bates: A Shaker's Journey*. Hanover: University Press of New England.

Mendel R (2013) *Closing Massachusetts' Training Schools: Reflections Forty Years Later*. Baltimore, MD: The Annie E. Casey Foundation.

Martin, B (1976) The Massachusetts Correctional System: Treatment as an ideology for control. *Crime and Social Justice* 6: 49–57.

Merton T and Pearson P (eds) (2011) *Seeking Paradise: The Spirit of the Shakers*. Maryknoll, NY: Orbis Books.

Miller MS (2007) *From Shaker Lands and Shaker Hands: A Survey of the Industries*. Hanover: University Press of New England.

Miller MS (2010) *Inspired Innovations: A Celebration of Shaker Ingenuity*. Hanover: University Press of New England.

Mollyneaux D (1987) *75 Years: An Informal History of Shaker Heights*. Shaker Heights: Shaker Heights Public Library.

Moran D (2014) Leaving behind the "Total Institution"? Teeth, transcarceral spaces and (re)inscription of the formerly incarcerated body. *Gender, Place and Culture* 21(1): 35–51.

National Park Service (2013) Shirley Shaker Village. *Shaker Historic Trail, National Register of Historic Places Travel Itinerary*. Available at: www.nps.gov/nr/travel/shaker/shi.htm.

Nicoletta J (1995) *The Architecture of the Shakers*. Woodstock, VT: Countryman Press.

Nicoletta J (2001) The gendering of order and disorder: Mother Ann Lee and Shaker architecture. *New England Quarterly* 74: 303–316.

Nicoletta J (2003) The architecture of control: Shaker dwelling houses and the reform movement in early nineteenth-century America. *Journal of the Society of Architectural Historians* 62(3): 352–387.

Paterwic SJ (2008) *Historical Dictionary of the Shakers*. Lanham, MD: The Scarecrow Press.

Patterson DW (2000) *The Shaker Spiritual*. Mineola, NY: Dover Publications.

Phaswana-Mafuya N and Haydam N (2005) Tourists' expectations and perceptions of the Robben Island Museum – A world heritage site. *Museum Management and Curatorship* 20: 149–169.

Promey S (1993) *Spiritual Spectacles: Vision and Image in Mid Nineteenth Century Shakerism*. Bloomington: Indiana University Press.

Rothman DJ (1971) *The Discovery of the Asylum: Social Order and Disorder in the New Republic*. Boston, MA: Little Brown.

Ruetalo V (2008) From penal institution to shopping mecca: The economics of memory and the case of Punta Carretas. *Cultural Critique* 68: 38–65.

Sasson D (2007) Dear Friend and Sister: Laura Langford-Holloway and the Shakers. *American Communal Studies Quarterly* 1: 4.

Scott S (2011) *Total Institutions and Reinvented Identities*. London: Palgrave Macmillan.

Skotnicki A (2000) *Religion and the Development of the American Penal System*. Lanham, MD: University Press of America.

Starbuck DR and Swank ST (1998) *A Shaker Family Album: Photographs from the Collection of Canterbury Shaker Village*. Hanover: University Press of New England.

Stein SJ (1992) *The Shaker Experience in America*. New Haven, CT: Yale University Press.

Stiles LA (2013) *Shaker "Great Barns" 1820s–1880s: Evolution of Shaker Dairy Barn Design and Its Relation to the Agricultural Press*. Clinton, NY: Richard W. Couper Press.

Strange C and Kempa M (2003) Shades of dark tourism: Alcatraz and Robben Island. *Annals of Tourism Research* 30(2): 386–405.

Swank S (1999) *Shaker Life, Art and Architecture*. New York: Abbeville Press.

Trompeter G (2012) New York's Craig Colony for Epileptics: Tracing the deepest roots of deinstitutionalization. *Middle States Geographer* 45: 76–83.

Trustees, Industrial School For Boys (1911) *Second Annual Report of the Trustees of the Industrial School for Boys at Shirley*. Boston, MA: Wright & Potter Printing.

Trustees, Massachusetts Training Schools (1916) *Fifth Annual Report of the Trustees of Massachusetts Training Schools*. Boston, MA: Wright & Potter Printing.

Wergland G (2006) *One Shaker Life: Isaac Newton Youngs, 1793–1865*. Amherst: University of Massachusetts Press.

Wergland G (2007) *Visiting the Shakers, 1789–1849: Watervliet, Hancock, Tyringham, New Lebanon*. Clinton, NY: Richard W. Couper Press.

Wergland G (2010) *Visiting the Shakers, 1850–1899: Watervliet, Hancock, Tyringham, New Lebanon*. Clinton, NY: Richard W. Couper Press.

Whitson RE (1983) *The Shakers: Two Centuries of Spiritual Reflection*. New York: Paulist Press.

8 Cartographies of affects

Undoing the prison in collective art by women prisoners[1]

Susana Draper

Introduction: historical geographies

This chapter looks at historical geographies of prisons through the relationships among temporality, architecture, and affects, by focusing on the ways in which collective works by prisoners configure past and present places of imprisonment. As research has shown, part of the anxiety surrounding *institutionalization* in the experience of imprisonment arises from the perceived effects of an environment promoting a system of machine-like repetition (Wahidin, 2006). Within this paradigm, the possibility of working collectively becomes difficult, even utopic, as prisoners are situated within a system built to promote solitude and lack of trust. However, as Sibley and van Hoven state (2009), it is important to search for mechanisms of the production of spatiality and temporality within the experience of imprisonment, instead of just looking at prisons from the gaze of the guards and builders of the system. In this chapter I analyze different forms in which imaginary cartographies of prison space and time are built *collectively* by prisoners. By focusing on collective rather than individual works related to the cartography of prison space, I aim to explore the ways in which the process of figuring a memory of the past and present of imprisonment can become a form of active and creative resistance.

To that end I engage with the process by which words and images emerge as zones that enable a polyphone of voices and meanings to spread, influencing the routine geography of confinement (made up of both legal and spatial discourses). I emphasize how these interventions become zones of struggle for the re-signification of the subjects (prisoners), the use of words (language), and images of space (prison walls), which lead us to rethink prisons as parts of a broader social and political network, and make us question the sense in which prisons came to be seen as natural solutions for dealing with inequality and violence (Davis, 2003; Martin and Mitchelson, 2008). That is, these interventions make us think about prisons in relation to different forms of inequality, including the ways in which neoliberal politics have both naturalized the criminalization of poverty and transformed prisons into a profitable industry (Gilmore, 1998/1999, 2007).

My project enables us to see a process which begins with the memory that enables the production of an historical geography and results in a form of collective empowerment.[2] In brief, I am interested in highlighting the process that goes *from* the practice of making a collective memory of imprisonment that enables the production of a historical geography *to* a form of collective empowerment that enables prisoners to reflect critically about imprisonment within a broader social network. I am interested in exploring two different processes and temporal frames in which such practices take place: (1) the collective act of rememoration of imprisonment in the past; and (2) the process of mapping imprisonment in the present of captivity.

First, I analyze a situation in which prison geography becomes historical *through the process of remembrance* when former women political prisoners from Córdoba (Argentina) created an artistic reconfiguration of space, when the prison from which they successfully escaped in the early 1970s closed in 2004 and opened later as both a cultural and commercial center. The struggle over the memory of the space is important in that it leads to questions about how memory may be linked to a transformed perception of prison space. Through this, a series of questions regarding gender stereotypes and memory of political resistance emerge in the former prisoners' performance, posing to us a question about what I call an *architecture of memory and affects*, working as a form of reaction and creative resistance to confinement. From this, I move to two different ways of re-imagining the space and time of confinement from inside prison, in works of poetry and mural painting carried out by current common prisoners while doing time at the Santa Martha de Acatitla facility in Iztapalapa, built in 2001 on the outskirts of Mexico City. I here articulate the role of the word in the women's collective workshops, as well as the visual narratives that led to the creation of four murals within the prison walls. The murals work as collective forms of memory building that produce an historical geography of the prison from within, transforming the walls into expressive mediums of a history of place. Through the painting, a praxis of historicization is at stake, one that creates a political sense of imprisonment and makes prisoners start to become politically active regarding their cases.

In the case of the women from Córdoba, memory of the past acts as a form of collective empowerment against the process of urban whitewashing. The former prison, now a commercial center, becomes a site of memory for them – continually changing as time passes. Pierre Nora (1989) conceptualizes *lieux de mémorie* as forms of materializing memory in place at the moment of "a turning point where consciousness of a break with the past is bound up with the sense that memory has been torn" (1989: 7). However, such memory is in a constant process of re-signification; as Elizabeth Jelin argues (2003), memory becomes the site of a struggle where different forms of relating with the past are enacted in a specific present. Each present articulates and re-invents a certain form of the past that works as a type of "rescue" of past elements in danger of being forgotten – the answer to the danger that is posed to a specific memory in the present; in this case, the memory of a collective form of escape from prison, which then becomes a collective attempt to escape from the market form. I am here following Walter Benjamin's idea that "history is not simply a science but also and not least *a form of remembrance*. What science has determined, remembrance can modify" (1999: 471). The memory that these women build may thus be seen in creative acts which are a response to the danger posed by the present situation in relation to their specific past struggles.

By analyzing these processes I aim to explore two different contexts and formats in which historical geography "takes place." One relates to a retrospective form, as former political prisoners build a memory of a place that had been significant in their political struggles of the past. The other relates to a form of historical construction within the present of captivity, where the task of giving a sense of space and time turns into the production of a political sense that attaches to the present and the past. In the first case, memory is a memory of politics that is remembered through the imaginary reconstruction of prison geography. In the second, memory becomes a powerful tool in creating a history in which the present takes place.

By posing these two temporalities and contexts, my goal is to relate the study of historical geographies of prison with the field of memory studies in Latin America, connecting past and present forms of imprisonment in order to think memory and territory together. Since the early 2000s, the issue of spaces and territories associated with the memory of state authoritarianism has been one of the major focal points for re-signification of dictatorships in the fields of history and the social sciences in Latin American studies. Based on Nora's notion of *lieux de mémoir* (1989) and on Maurice Halbawchs' collective memory processes (1992), Jelin's pioneering studies on memory works in post-dictatorship situations (2003) have foregrounded the relationship between territoriality and memory. Studies on politics of memory in Latin America have focused mostly on the processes of rememoration of the brutal military dictatorships in the 1970s and 1980s, placing memory as a form of struggle through which a complex re-creation of places of confinement and disappearance takes place. In this sense, memory emerged as a practice through which the past was constantly re-signified, thus opening up a field of contestation to the official demands to move on and forget about the past.

As a practice of signification of the past, memory becomes the site of a struggle where contested versions emerge and pose connections to the present.[3] However, within the field of memory studies in Latin American post-dictatorship contexts, something that remains unexplored is the possibility of carrying out an analysis of the different forms that the relation between territory and memory take in the *present* of imprisonment. That is to say, the practice of working collectively in the mapping of prison space becomes also a form of building memory and, therefore, of transforming the present time of imprisonment perceptible and figurable as historical time; i.e., a practice that allows the present to become critical, thus interrupting the machine-like repetition of a homogeneous time. As Wahidin (2006) argues, the organization of time in prison "seeks to deny the society of captives (Sykes 1958) the capacity to create meaningful and symbolic relations with prison-time and external time in the free society." By approaching different forms of temporal signification of past and present, I aim to connect the struggles for the re-creation of a memory of the past with the practice of signification of the present. It is also a way of thinking about the multiple forms that time acquires in the prison (Wahidin, 2006), and, building upon the possibility of approaching space through time, decentralizing the primacy of time in the approach to prison geographies (Moran, 2012). At the same time, I have chosen to focus only on works by women because there has been a problematic lack of attention given to the situation and symbolic struggles of Latin American women prisoners in both past and present times (Azaola, 2003: 92; Comisión, 2003: 9–10; Rodríguez, 2003).

Escaping the invisible prison of the present

The Buen Pastor prison in Córdoba (Argentina) was transformed into a cultural and commercial "Paseo" and memorialistic culture played an important role in the imaginary of the new, camouflaged commercial center. Like Punta Carretas prison in Montevideo (Uruguay), the Buen Pastor prison and chapel opened at the beginning of the twentieth century (in 1906),[4] and became an important detention center for women in the 1960s and 1970s. Closed as a prison in 2004, it was made into a cultural, commercial, and recreation center as well as a historical site called the Paseo del Buen Pastor. Although the goal of the transformation was to preserve the memory of the place while adjusting it to the commercial necessities of the city, the whole complex of prison cells was demolished and replaced by commercial lots. Buen Pastor prison was a key center for the detention of female political dissidents in the early 1970s. It became a symbol of resistance following the successful escape of 26 prisoners in 1975, in the lead-up to the coup when a state of emergency was declared in the province (*medidas de pronta seguridad*). The fact that Buen Pastor was associated, like Punta Carretas, with a collective escape and was subsequently turned into a camouflaged shopping and memory site ties it to the issue of the new economy of memory and the search for other forms of escape. The paradox here is that an escape also entails the possibility of a different form of administering remembrance and forgetfulness on behalf of both the state and the market.[5]

The female ex-prisoners of Buen Pastor gathered and initiated various actions in order to contest the form in which the transformation of the place was being carried out. A key issue was the renovation's erasure of the window of the escape, understood to be part of a process of invisibilization of acts of freedom. At the same time, the window worked as a form of border that was crossed at the moment of the escape. After a long struggle and working in collaboration with the Archivo Provincial – the provincial human rights archive – the women were able to situate the window bars back onto the site, thus re-engraving the escape on the memorial place, in effect differently signifying the newly renovated scene. This act was important because it contested the way in which certain forms of institutionalization of memory neutralize or even deny political content.

The architectural transformation resulted in the emptying-out or plain destruction of a space that was crucial to people's memory of the prison – including the shape of the cells and the structure of the cell block. The creation of this vacuum angered some of the surviving detainees, who protested the erasures despite their fear of repercussions (their protest took place at the moment in which anti-terrorist legislation was being discussed). Some ex-detainees attended the opening event and took the floor, interrupting the scheduled programming. They decorated the barriers protecting the Paseo with ribbons bearing the names of the prisoners who were killed or disappeared after the escape, reciting their names and chorusing "presente ahora y siempre" ("present now and always"). It consti-tuted what Mariana Tello (2010) calls a counter-act to the planned and scheduled event, and it posed a series of questions regarding the conflictual role of memory vis-à-vis politically framed narratives of the past.

The counter-event staged a different form of connection between past and present, inside and outside of prison, because, by placing ribbons with the names of the dead women who had escaped on the barriers, the survivors were paying homage to the disappeared, seizing on the poetics of the escape as their main trope of resistance to museification. The protest becomes even more interesting when read alongside a video clip created by the authorities to represent the past and present of the place as a linear continuity – from the dark, old place of con-tainment and repression to one of free consumption and modernization. In the video, prisoners are objectified in typical photographs depicting them in line, being registered at the entrance, suggesting that opening the former prison as a commercial center is tantamount to redeeming its dark past into a modernized space of consumption and free gathering. The lines of prisoners, the clip sug-gests, have now turned into lines of consumers using their freedom to enter the place. However, the video was in turn counter-acted by a documentary whose title points to the erased parts of the past: *Buen Pastor: Una fuga de mujeres* [*Buen Pastor: A Women's Escape*] (Torres and Herrera, 2010).

Co-produced by Cine El Calefón and a group of former political prisoners and participants in the 1975 escape, the documentary starts with two images of passage: first, the narration dramatized by the video clip in which the successful opening of the Paseo implied the passage from the dark history of the place to the new realm of consumerist freedom; and second, a kaleidoscope of prison memories that focus on the *window* that was used for the escape three decades earlier. Here, a different meaning of the word "Paseo" surfaces, referring to the "walk" of former prisoners who try to locate themselves in the transformed place in order to recreate the layout of the prison where they had once lived, and an imaginary map of the path leading to the escape. Recording former inmates in the process of trying to locate themselves, and to identify the location of the window from where they escaped, the documentary shows a dialogue of com-peting memories as women argue about their memories regarding the layout of the place, the location of their beds, the bathroom, etc.

As if to teach viewers another itinerary of Buen Pastor, a singular act takes place in the documentary while people are eating and shopping. Temporal layers mix as the women progressively transform the place into a site of struggle in which dif-ferent diagrams of remembrance take place. Remembering the resistance of the past, the polyphony of voices reinventing history in the present is an act of resistance to the present physiognomy of the place, not only as a space for consumption but also as one that confines the female inmates' history to the easy stereotype of victim and suffering mother. The new, clean, and colourful center becomes the backdrop for a collage that tunnels to a different narration in which the women's struggles in prison and the preparations for the escape encourage them to start escaping from the commercial center. This reaches a climax when one of the former prisoners being filmed, Cristina Salvarezza, starts writing graffiti on the walls and pillars of the former prison. Remarkably, all her graffiti refer to women inmates who disappeared following the escape, and after she inscribes the pillars with phrases such as "Here, Tota used to sing and Mariana danced," Salvarezza starts to sing and dance the songs she remembers, creating a strange point of temporal crossing on camera where the memories of resistance in prison encounter a present in which resistance is being held captive in the commercial center. The idea of "making" the territory with memories of common activities in everyday prison life, ones full of affective components (such as joy, sadness, hope), creates an atmosphere that estranges the space and its borders. That is, by means of performances within the documentary, there is an interesting form of re-imagining the geography of prison space. As former prisoners remember the prison space through different activities (eating, gathering, dancing, singing, planning the escape), one sees that the erased window that marked the border between in and out of prison works as the organizer of the new imaginary map, but also turns the outside in, as the memory focuses on the way in which the bars were challenged by the escape.

By trying to locate where the "window" formed as they stepped onto the outside, the survivors introduce a kaleidoscope of past images that resist easy categorization as prisoners, mothers, militants, or suffering victims, creating a different poetic. Such resistance to the stereotypes seems to be marked by the figure of the escape that seems to move from the act of remembering a fact to different signifying chains that the former prisoners are destabilizing and trespassing. Thus, by bringing back the escape to a signifying chain that affixed easy labels to them, the ex-inmates create a constellation of images of an irrepressible past of affection and struggle. In Buen Pastor, the marketing of memory that confined the past to a set of comfortable stereotypes made former prisoners transform the site of memory into the place of a struggle over meaning. I would even go further and say that this is a struggle which attempts to keep open a space for the possible re-signification of the political in non-stereotyped ways. As I mentioned above, the insistence on bringing up the topic of escape was a struggle over the place and over the politics of memory itself. Following these protests, the former escapees successfully lobbied for the Archivo to allow them to place a plaque remembering the escape, thus opening up a form of historicization that does not limit their imprisonment to a passive form of suffering, and recovering the search for freedom that led to their escape. The struggle to add other meanings continues.

Once they had secured authorization, the ex-prisoners also started to *mark the site*, configuring an interesting pastiche in which the restaurants and stores were interrupted by inscriptions on the ground remembering the disappeared women.

The most prominent figure within this is the installation of the window from where the prisoners are believed to have escaped. This creates an interesting image in which past and present collide in a silent dialogue: the window, working as the border (the bars that were surpassed), located close to the tables of a fancy restaurant, marks the border of the former prison, bringing to the present a trigger to remember and re-imagine those parts of the site that have been erased: the cells, the patio. The window is inserted as if coming from a different time, grey and old, and coexists with the colors of the market, leaving open the metaphor of an inside of prison that is now the place for consumerist freedom. The idea of taking the notion of the escape as the center around which a memory of the space and time of confinement was recreated is telling, as it aims at destabilizing the segmented and rigid organization of prisons, and *othering* the understanding of material space and its temporal signification. It brings the past in the form of an *architecture of affects* that endows a white-washed space with a historical dimension.

Painting the memory of present imprisonment

The work of memory which the former political prisoners carried out through the collective actions that I discuss above belongs to a struggle for a re-signification of the past, contesting the transformation of prison space into a site of neoliberal consumption. A question that permeates the whole set of actions involves the figure of the escape – namely what an escape would mean within the present of that transformed prison. In a way, the work of memory carried out by the prisoners itself becomes a memory of politics: the memory of a collective process of escape. The process of building an imaginary cartography of the former prison at the moment of its transformation becomes a site for the empowerment of the collective of survivors. They struggled until the window from which they had escaped was reinserted into the middle of the site, thus interrupting the landscape of consumption. The possibility of building an architecture of affects seems to become the site for a new escape, where the prisoners protest the invisibility of the prison in the consumerist society.

When analyzing this case, I began to search for the possibility of a similar process within the situation of current imprisonment – not as a memory that builds and contests a cartography of prison space, or time in the past tense, but rather within the present state of imprisonment. In a way, this would make us think about memory in a different context and temporal span, with the center not a past of collective actions to be remembered, but rather the memory of a present time (i.e., the temporal horizon of a present that seems to be atemporal in the experience of imprisonment). The question at stake here is that of the possibility of making the time of imprisonment historical – charged with historical meaning produced by the prisoners themselves, as a form of struggle with temporal dimensions that the prison system imposes to life spent inside.

This is a key problem, since it relates to the possibility of looking at the temporal dimension of carceral geography, often relegated to secondary consideration (Moran, 2012). At the same time, I am interested in cases in which prisoners become able to work on *collectively signifying time* in prison, and therefore beyond the form of the individual experience of temporal perception. I analyze the relations between time and prison experience explored in depth by Wahidin (2006), but I focus specifically on prisoners' collective creative works. By centering on the collective, I am interested in seeing how the process of signification of space and time can become relevant in forging a sense of the future.

In the creation of literature and art, prisoners express forms of resistance and protest that aim to reconceptualize and re-historicize the space of confinement. That is, the process of thinking about the space and intervening in it becomes a form of signifying the temporal dimension of imprisonment and of translating the abstract time of a sentence onto the lived experience. This has been exemplified in an artistic project held at the Mexican women's prison facility of Santa Martha the Acatitla, Mexico, consisting of a series of literary workshops that led to the creation of collective visual narrations on the prison walls.

From 2008 to the present, women working at the Program in Gender Studies at the National Autonomous University of Mexico (UNAM) started a workshop with a group of common prisoners at the women's facility of Santa Martha de Acatitla. What began as a reading workshop led to the creation of a mural inside of the prison, producing a prison research-action group project called "Women in spiral: system of justice, gender perspective and pedagogies of resistance" [*Mujeres en espiral*: sistema de justicia, perspectiva de género y pedagogías en resistencia]. One can notice certain forms of empowerment that the creation of visual narratives evoked in the women prisoners, as one mural led to another, with a total of four wall interventions in the prison. The last one led to the creation of a legal clinic inside the prison, Clínica de Justicia y Género Marisela Escobedo, founded with the support of the School of Law of the university (UNAM) and the Human Rights Commission of Mexico City. It is interesting to see how the creation of visual and verbal narratives about the situation of imprisonment, while actively intervening in the space of the prison itself (painting the walls), endowed the time of imprisonment with a historical component in which the murals became a form of collective memory. The textual and visual work led to the creation of a legal clinic in which prisoners can learn about their processes and struggle for them. This suggests a movement from solely space intervention and temporal signification to a form of legal empowerment. By analyzing some of the words and images reproduced in the volume that explains this process, in the book *Pintar los muros, deshacer la cárcel* [*Painting the Walls, Undoing the Prison*], I would like to focus on the ways in which creative interventions in the prison space unveil a complex perspective of the prison system. I also explore how they reconfigure the materiality of time in relation to the idea of the walls that divide the inside and outside, the unfree and the free.

The aforementioned project began with a reading and writing workshop, where prisoners familiarized themselves with different texts written by Gloria Anzaldúa, Rosario Castellano, Elena Garro, and José Revueltas, among others. These readings worked as an instigator for the creation of brief narratives training the senses, in order to start imagining the visual form the murals would take. It is interesting to note that the contributors posed the idea of painting murals as a form of "take-over" (la toma) of prison space, which means the possibility of moving from experiencing the prison as something that objectifies women (a space that encloses them, depriving them of agency) to the possibility of becoming agents in inhabiting the space, able to send messages to the outside world. The titles of the four murals they have created so far are telling: "The scream" [El grito]; "Force, time, and hope" [Fuerza, tiempo y esperanza]; "Paths and forms of freedom" [Caminos y formas de la libertad]; and "Collective action for justice" [Acción colectiva por la justicia].

Marisa Belausteguigoitia, director of the program on gender studies at the university (UNAM), states that the initial goal of the project was to "make visible and to interrupt the walls that make imprisoned women feel reduced." By painting murals, the idea was "to shed light on the subjective, educative and juridical aspects" at stake in the process of incarceration (Belausteguigoitia, 2013: 12, my translation). Within this framework, the walls that are essential for the delimitation of the prison as a social function (the inside-outside) were transformed into a series of subtle, invisible actions, tackled through the creation of visual narratives that sought to "undo" the prison. In prison jargon, "hacer cárcel" – *to do the prison* – means to shrink, to become small, to wall oneself in a system of suspicion and obedience (Belausteguigoitia, 2013: 13). Playing with this jargon, the idea of a collective creation of visual narratives refers to the act of "undoing" the prison, the process of building a subjective and collective awareness of the prison as a situation, and of the walls as something more than mere spatial delimitation. At the same time, the focus on gender relates to the creation of a form of awareness of an invisible, naturalized system of abuses. Thus, the process of "undoing" the prison immediately connected with the idea of undoing the dominant mythology of the "good and docile woman," which functions as a mysogynistic paradigm that ends up walling women inside (Almeda, 2002; Antony, 2003; Azaola, 2003; Lagarde, 1990, 1996; Moncayo Gómez, 2014).

The first notes taken for the production of a mural relate to the possibility of breaking up silence and opening up language. The mapping of the space starts with a reflection on silence, on the lack of discourse, and reflection on imprisonment by the prisoners themselves:

How much silence
has accumulated here?
And inside women?
What is silence?
Silence is the void of words,
of evidence that would give an account
of who I am, of what I am here,
what is my name, who cares about me,
what I eat, what takes my sleep away,

of what color and at what temperature,
when is water available for us, here.
Silence is to not count
(to not being taken into account
and not giving account of that).
Silence is passion, emotion,
it is to look for the word, the image,

(Marisa, in Belausteguigoitia, 2013: 30)[6]

The reflection immediately moves on to the connection between space, gaze, time, and architecture. The architecture of this prison has a complicated form of organizing the gaze: it is designed in a way that makes it impossible for prisoners to see beyond ten meters; that is, the space is designed so that, wherever you look, you will reach a bar, a guard, a wall. Characterized as both labyrinth and panopticon, the gaze of the women inside is constantly controlled and restricted. The optical experience and perception throughout the time of captivity makes the notion of a horizon physically impossible (Belausteguigoitia, 2013: 66). This is, of course, transposed onto a series of feelings reiterated in the women's writings: paralysis, lack of horizon, lack of sight, lack of futurity. The space organized the optical dimension of life, and both compose a temporal sense without horizons beyond the present. The impossibility of looking beyond ten meters and visualizing a horizon inhibits thinking of a future, or a long-term history – as Dirsuweit argues, the re-signification of space becomes a form of resistance to "spatialised structures of control" (1999: 73). I would add that such a form of re-signification becomes a trigger of other forms of experiencing temporality, one that differs from the sense of a temporal anesthesia – the dulling of the possibility of sensing a past and a future. As one prisoner (Ethel) states: "Pain dumb you in the prison, it wraps you up as a time capsule where you don't care about anything and where *you don't remember anything*" (in Belausteguigoitia, 2013: 78–79, emphasis added). It is interesting that the prisoner connects such organization of the visual field with a sense of time, a "capsule" without memory and history, a present without any form of past or future.

Time and language emerge as an obsession in the notes taken in order to create the first mural, the ability to take the wall seen as a possibility of "naming" beyond binaries:

How do you feel? Good or bad? Are you happy or sad?... *There is no learning process that would teach us to name what we feel beyond these two poles.* To be able to name an emotion and to embody it and also to give it color.

(Claudia, in Belausteguigoitia, 2013: 33)

In the creation of the first mural, "The Scream," the women were able to express affects not subsumable under common binaries (outside–inside, free–unfree), which commonly paralyze women inside. If the point of departure is the prison – the state of being imprisoned (walled) – the problem faced by the women who decided to start building outside of it relates to an impossibility of *finding a language (in both words and images)* to express more ambivalent emotions. It is interesting that in order to collectively process the ideas that would be "figured" in the mural, an intense work on language had to occur, with the architectonic of prison life having penetrated forms of saying and seeing.

Abuse, anger, loss, pain, *but mostly paralysis*, are the feelings that consistently appear in the book's passages, from the moment of starting the workshops for "The Scream." The mural was painted on a spiral staircase; the place chosen is, itself, a zone of passage: the stairs leading to the visiting room when going downstairs, and to the area where prisoners are sent before being freed when going upstairs. It is interesting that the action of taking over this place for the first mural and painting its oblique walls implies the possibility of meditating about passaging, the space-between, which in itself makes conversations about ambivalence in feelings and situations possible. At the same time, it is a difficult site to think about, as it connects to many complicated affects linked to the outside. The series of notes and texts that compose the process of building the first visual narrative are mostly related to different emotions – fear, anger, anxiety, solitude – and the search for a language to express them in words and colors. They are a form of breaking silence, without falling into the usual dichotomies in which prisoners feel doubly incarcerated; in a way, the act of collectively transforming prison walls into murals is, in itself, a form of transforming the power of authority expressed in prison architecture. It is a form of *spacing*, understood in different ways: opening up a signifying process within the most rigid delimitation of the prison as such, creating a site in-between that problematizes the rigid categories of inside–outside, while expressing incommensurable forms of experiencing temporality, spatiality, everyday life.

Taking Foucault's (1996) idea of the prison diagram as an optical and architectural system, one could say that the murals perform an intervention that interrupts the eyes imposed by the dominant system of imprisonment, in turn producing a different perception of the architecture itself. The exploration of the role of vision, the sight, and the possibility of creating a temporal horizon plays a key role in the creation of the first visual narrative. The opposition is clear. On the one hand, there exists the gray space where long-distance vision is impossible and where women are tracked by cameras in all the corridors and prison space. At the same time, the idea of the workshop leading to the painting was to enable the prisoners to contest the condemnation of being paralyzed "objects" of surveillance, "able to look beyond, create horizons and perspectives that broaden their visual, spatial, and subjective horizons" (Belausteguigoitia, 2013: 66). The drawings painted in the mural play with the idea of the sea, horizons, green vegetables, colorful flowers, children, and lots of women.

As Baer shows in his analysis of visual imprints on prisons, "a mark on the wall might transform a space within a prison to give it a different meaning" (Baer, 2005: 211). Little by little, the words and colors produce something important regarding the way in which time is sensed inside. For instance, the first murals painted on the staircases bestowed the space with a different meaning: affects (the way to the visit); memory (remembering the emotions when walking to see people coming from the outside); and also freedom (of being able to intervene in a space where they have been objectified as passive subjects). One muralist (Polo) writes: "Right now [*ahorita*], what we are doing is to take the wall as a notebook" (Belausteguigoitia, 2013: 35). In this, one can see the idea of a space that can become reflexive, the material for a critical action instead of feeling the space as a mere imposition. What is important here is that the intervention in the space where they spend days, weeks, months, years, decades, starts to be felt as *a source of action* – the action of undoing it, resisting while opening up a different form of relating among themselves, thereby destroying the isolation and the suspicion that the prison builds in them. This also established them as a group. The name, "Women in a spiral," suggests the sense of movement, process, the idea of sharing affects, bodies, freedom, as well as the idea of their lives as a place they need to start thinking about within this other horizon.

It is interesting to point out that the figure of the spiral also emerges as a powerful component in the notes prisoners kept while working on the figures in the mural. In essence, they passed through different ambivalent feelings and affects, looking for ways to relate the meaning of their gathering and talking, so as to paint and re-signify their relation to the space. The spiral is also used as a form of walking, a trope that comes from the zapatista imaginary, employed in numerous zapatistas postcards. I think that the spiral emerges as a form of breaking the circle, thus posing the possibility of adopting an active role within a situation that was, until then, felt as a destiny, a case closed.

> To walk in a spiral through space,
> fast and slow,
> stopping and chatting with our friends.
> To move in spiral,...
> What is force?
> To walk over the space.
> What is hope?
> All fainting,
> having fun,
> a sense of freedom.
>
> (Arelhí, in Belausteguigoitia, 2013: 140–141)

This passage suggests a spiral in other forms: paths, walking together, sharing affective instances that also demonstrate the possibility of sharing the joy of becoming a group able to determine itself. In a way, after "The Scream," the themes that emerge for the second mural are more related to addressing a different set of problems, focusing on the perception of time and incommensurability.

The prison is emplaced now as an act of sharing the singularity of affects that cannot be *measured*. Thus, the second mural, "Force, Time and Hope," moves from the individual feelings gathered in the notebooks that led to "The Scream" to a search for an undoing of the prison by themselves, as *a group*. Fifty-two women who had not spoken to one another before because of their status divisions (the sentenced and the ones awaiting their sentences) participated in this group, forging a collaboration that had never before taken place in that prison. The discussions which took place in preparation for the mural led to the prisoners working together to problematize the perception of time across these divisions.

While going through the passages, words, and images, many questions arose: What is the time of the prison? Is it the time of the individual sentence? How is the measure of the time of a sentence different from the measure of time outside of the prison? How is social time composed? Who are the subjects able to live in a common social temporality? Can there be a *common sense of time* in a prison where each prisoner counts time according to her specific sentence? Can a collective form of temporalization take place through the process of creating visual narratives that change the cartography of imprisonment? Their works are full of drawings with clocks. One shows a woman carrying a huge sand-clock chained to her wrist on her back. In the mural there are plenty of sand-clocks, each containing stories, playing with the struggle with a sense of "dead-time" [tiempo muerto] as a temporal form of time perception in prison. The drawings of small sand-clocks with people, faces, hands, are super-posed to a back-drawing that has bigger clocks with small sailboats floating on a river. The drawings of different clocks are accompanied by many arms that are encroached, one among others, and the notes taken on this part of the mural are interesting, as they play with different, non-homogeneous forms of perceiving, feeling, and measuring time: "There's a lot of people who thank God for one more day. We thank for one less. On day less in prison" (Ethel, in Belausteguigoitia, 2013: 152–153); or "Here we have all the time in the world" (Natacha, in Belausteguigoitia, 2013: 160–161).

Through the work on the temporal signification of imprisonment, a sense of futurity emerges in connection to the third mural, "Paths and Forms of Freedom," which involves a change from the spiral to a plain, straight wall. For the first murals, the space was a passageway in terms of ascending and descending, but now the configuration of a visual narrative is connected to the inside–outside of prison. In a wall seven meters in height, they draw once again the forest with lots of vegetation and green (signaling the lack of nature in the life spent in prison). New figures emerge: pyramids, an Aztec sun, a path of kisses [Callejón del beso], and a big sea. A note on this mural says:

I draw a sailing boat, to make it take me to the other side of the world/
That's why I am looking towards the sea, towards the exit/
To see inside. To go far away in this boat.

(Lulú, in Belausteguigoitia, 2013: 185–186)

We see women dressed up to go to the beach, sitting in front of the sea, waving their hands hello or goodbye to the horizon. It is also interesting that the sun giving light to the landscape is borrowed from the Aztec imaginary, and it is waving a long tongue, which is a sign of laughter, of making a joke, but also one of disobedience. "Sacar la lengua" is used in Spanish to mock authority in a laughing gesture. This, in a way, connects to the work seen in the fourth mural, called "Collective Action for Justice," located on the wall closest to the street. At the center of this mural is a figure of Law (the penal code), in the form of an almost-closed door. The mural conveys the situation of illiterate women before the law, which led immediately to the creation of legal clinics for women in the prison (above). The generalized feeling was the lack of understanding of the language that was being used to keep them imprisoned: "The documents of the lawyers fly away and they do not have order or time or space … they are flying and they do not say anything for us" (Aida, in Belausteguigoitia, 2013: 220–221).

The whole mural plays with a solar system, consisting of many figures of women standing with erased faces. The passages here say: "We will push the wall" (Marisa, in Belausteguigoitia, 2013: 208); "That free element, that they cannot take away from you, that part without bars, without barbed wire, without razors" (Belausteguigoitia, 2013: 218–219). One of the final murals portrays time in prison as inverse to profit, calculating the money needed to attain freedom: "They gave me 10 years for 7 thousand pesos, 1,000 a year and I can go out in 7" (Chuy, in Belausteguigoitia, 2013: 225); "Door, they call you freedom" (Liz, in Belausteguigoitia, 2013: 231). These last words make us reflect on the role of numbers and space in relation to freedom. In a way, the process started by the murals demonstrates the intimate connection between the prison space and the way in which it affects and determines the time of the lives spent inside of it. By intervening in the walls themselves, the questions that arise through these projects point to the overarching question of the form in which the prison architecture and optics of power may be "inhabited."

From imprisonment to historical agency

What does the work on the walls mean in terms of a collective process of *agency*? How does it relate to the process of giving an historical sense to the space? Walls relate to imprisonment as an everyday, imperceptible act that operates on bodies, feelings, affects. This is what the prison jargon means by "to make prison" [hacer la cárcel] in terms of the direct form of effectuation of the prison system through the organization of space, the architectural and optical system operating on the bodies contained there. As Wendy Brown (2010) states, walls in themselves do not *refer* to or *mean* anything; however, they "are potent organizers of human psychic landscapes generative of cultural and political identities"; that is, "they can become discursive statements themselves, and they are crucial to the organization of power in and through space" (2010: 74). Seen from the architectural standpoint of the dominant power system that structures the prison, walls are essential for any new production of imprisonment (i.e., to make walls speak in a different way).

Walls enclose prisoners. They are the first material dividers in the spatializa-tion of "being-inside" for the unfree. Therefore, the act of making the walls expressive in a different form – making them *speak* another language than that of the architecture of authority – relates to the possibility of transforming the form of imprisonment in a subtle, but still potent, way. I do not want to say that by painting on them prisoners are "free," as the prison is still at work in this very process. What I would like to point out is how the process of taking over such a relevant spatial mark *is doing something* to the signifying regime of prison life; here the prisoners actively intervene. Undoing the material spaces through which structures of power stabilize and reproduce themselves implies then a form of erasing and redrawing the rigid divisions that result in the prison system, the social diagram of which it is a part, and the conceptualization of social freedom that is therein at stake. In works by prisoners, the main issue questioned, sus-pended, or interrupted is the *signification of freedom* at stake in those spatio-temporal measures. A minor intervention seems to relate to the problematization of the notion of borders that one usually takes for granted, such as: the question of where the prison starts and where it ends; the time of freedom in situations of social inequality; the production of the ideal subject of social freedom; and from where we can "read" these demarcations.[7]

In a way, I see the above works as effectuating a constructive critique of the social diagram in which the prison is seen as a solution, and from there it is also making prisoners perceive their way of inhabiting not only space but also the *time* of imprisonment differently. Becoming a collective in the process of deter-mining their lives within is a way to begin a process of undoing the imprison-ment within the prison: the isolation, the sense of solitude that produces impotency. This is something that the second mural depicts in a singular form, with all the sand-clocks full of stories, akin to a Russian doll structure where one clock contains many other clocks – a complex system of multiple temporalities that counteract the homogenization of the time and life of prisoners, and the reduction of their horizon onto the passive form of waiting.

There is then a transformation of the role of art at stake here, as it becomes a *material collective practice of transformation* of inhabitation: it transforms the dead-end idea of the wall onto an open horizon, and the horizon is named "path" and "freedom." In a place that is designed to effectuate an impossible gaze beyond (what I explained above regarding the impossibility of seeing something other than a wall or a guard within ten meters), such transformation of dead-end into horizon also suggests a possible transformation of the perception of impris-oned lives. In Spanish, the sentence has two possible referents: *sentencia* or *condena*. The latter is highly relevant, as it implies the notion of a life devoid of freedom – everything is predetermined: poverty, abuse, criminality, poverty, exclusion, abuse. This suggests that time as cyclical. However, by breaking the cycle and transforming it into a spiral a different possibility emerges: a trans-formation of *process* (i.e., *the process of being able to move to a different place*). This implies a transformation of the perception of time, space, and freedom, re-signifying the space from a new perspective. In this re-signifying practice, the final mural visually addressed the problem of the Law and the system of "Justice" through spatial forms (Law painted as an almost shut door, prisoners with erased faces), which became an active form of dealing with the structural injustice of a law that did not speak for them. The creation of an in-prison legal and gender clinic is then a form of enacting the theme of the mural, a practical continuation of the process represented within it.

I have analyzed how mural painting, by moving to the realm of physical space in the prison, and the bodies that spend their lives *in* there, became a form of questioning the prison; that is, instead of adapting art to the prison form, the take-over of the space of prison in order to create a mural creates a process of undoing usually unquestioned power relations. A simple, but potent, form of telling, the murals challenge the entire control of the bodies by means of permanent surveillance and lack of sight beyond ten meters. This is thus a form of questioning the ways in which architecture affects bodies, and makes bodies react against it, communicating a sense of collective action from within to the outside. At stake is a collective reflection on the relation between architecture and bodies, posing the sense of freedom (a form of walking together) intimately tied to a material practice, a transformation of the space where life takes place. The challenge for us is to envision how these murals work to create a sense of time and justice that takes place not only within the prison walls but also in those that permeate the broader social world, its "constitution," its penal codes.

The interventions that I analyzed here – of former political prisoners at the Buen Pastor commercial center and of current prisoners at the Santa Martha de Acatitla facility – engage forms of memory that differently compose prison space. In a way, the role the window plays in the re-creation of the memory of the past in Buen Pastor relates to the figure of the wall in the construction of a different temporal sense in the case of Santa Martha. Both act as central points from which space is used in a novel way. In the former case, this occurs against the whitewashing of memory produced by the market through the recycling and transformation of prisons; in the latter, it occurs against the erasure of a temporal dimension in the prisoners' lives enacted by the architecture through its organization of space and vision. At stake in both is a form of reclaiming the act of historization of carceral space from the perspective of those who had or have been imprisoned. The vacuum of time that the consumerist time enacts in the case of Buen Pastor correlates with the sense of a capsule of time in Santa Martha; the possibility of collectively organizing an alternative temporal dimension is a key component of both cases, relating to the possibility of building a cartography made of affects and memories.

Notes

1 I would like to thank Dominque Moran and Karen Morin for their invaluable comments and feedback in the process of writing this chapter.
2 For an analysis of the necessity of studying the function of time within the cartography of imprisonment, see Moran (2012). For an analysis of the multiple forms that time acquires in the experience of imprisonment and in relation to the body, see Wahidin (2002, 2006; Wahidin and Tate, 2005).

3　I here follow Catela's (2007) distinction between processes of long- and short-term memory, as it allows us to distinguish the memory of struggles against the most recent dictatorships from the memory of other past struggles for social justice.

4　The opening of Buen Pastor may be seen as part of what Salvatore and Aguirre (1996, 2001) historized as a process of modernization based on the ideal of social regeneration at the end of the nineteenth and beginning of the twentieth century in Latin America.

5　For a detailed analysis of this prison transformation within a broader context of architectural reconfiguration in post-dictatorship Latin America, see Draper (2012).

6　All the quotes come from the book *Pintar los muros*, and all translations are my own.

7　In a recent book, Jill Stoner proposes the idea of thinking about minor architectures as "opportunistic events in response to latent but powerful desires to *undo structures of power;* and as such, minor architectures are precisely (if perversely) concerned with the privilege and circumstances of major architecture, the architecture of State and economic authority" (2012: 7). We could extend the notion of major architecture as an architecture of authority, following Richard Ross' (2007) work on photography.

References

Almeda E (2002) *Corregir y castigar. El ayer y hoy de las cárceles de mujeres.* Barcelona: Ediciones Ballaterra.

Antony C (2003) Panorama de la situación de las mujeres privadas de libertad en América Latina desde una perspectiva de género. In Comisión de Derechos Humanos del Distrito Federal (ed) *Violencia Contra las Mujeres Privadas de Libertad en América Latina.* Washington, DC: Due Process of Law Foundation, 75–90.

Azaola E (2003) Género y justicia penal en México. In Comisión de Derechos Humanos del Distrito Federal (ed) *Violencia Contra las Mujeres Privadas de Libertad en América Latina.* Washington, DC: Due Process of Law Foundation, 91–108.

Baer L (2005) Visual imprints on the prison landscape: A study on the decorations in prison cells. *Tijdschift voor Economische en Sociale Geografie* 96(2): 209–217.

Belausteguigoitia M (2013) *Pintar los muros. Deshacer la cárcel.* Mexico DF: PUEG-UNAM.

Benjamin W (1999) *The Arcades Project.* Cambridge, MA: Harvard University Press.

Brown W (2010) *Walled States, Waning Sovereignty.* New York: Zone Books.

Catela Da Silva L (2007) Poder Local y violencia: Memorias de la represión en el noroeste argentino. In Isla A (ed) *Inseguridad y violencia en el Cono Sur.* Buenos Aires: Paidós, 211–227.

Comisión de Derechos Humanos del Distrito Federal (2013) Presentación. In Comisión de Derechos Humanos del Distrito Federal (ed) *Violencia Contra las Mujeres Privadas de Libertad en América Latina.* Washington, DC: Due Process of Law Foundation, 9–11.

Davis A (2003) *Are Prisons Obsolete?* New York: Open Media.

Dirsuweit T (1999) Carceral spaces in South Africa: A case study of institutional power, sexuality and transgression in a women's prison. *Geoforum* 30: 71–83.

Draper S (2012) *Afterlives of Confinement: Spatial Transitions in Post-Dictatorship Latin America.* Pittsburgh: University of Pittsburgh Press.

Foucault M (1996) *Discipline and Punish: The Birth of the Prison.* New York: Random House.

Gilmore RW (1998/1999) Globalisation and US prison growth: From military Keynesianism to post-Keynesian militarism. *Race and Class* 40(2/3): 171–188.

Gilmore RW (2007) *Golden Gulag: Prisons, Surplus, Crisis, and Opposition in Globalizing California.* Berkeley: University of California Press.

Halbawchs M (1992) *On Collective Memory*. Chicago, IL: Chicago University Press.

Jelin E (2003) *State Repression and the Labors of Memory*. Minneapolis: University of Minnesota Press.

Lagarde M (1990) *Los cautiverios de las Mujeres: Madresposas, monjas, putas, presas y locas*. México: UNAM.

Lagarde M (1996) *Género y feminismo: Desarrollo humano y democreacia*. Madrid: Horas y horas.

Martin L and Mitchelson M (2008) Geographies of detention and imprisonment: Interrogating spatial practices of confinement, discipline, law, and state power. *Geography Compass* 31(1): 459–477.

Moncayo Gómez M (2014) Mujeres en prisión, los casos de Santa Martha Acatitla. Available at: http://clepso.flacso.edu.mx/sites/default/files/clepso.2014_eje_6_moncayo.

Moran D (2012) "Doing time" in carceral space: Timespace and carceral geography. *Geografiska Annaler. Series B, Human Geography* 94(4): 305–316.

Nora P (1989) Between memory and history: Les Lieux de Mémoire. *Representations* 26: 7–24.

Rodríguez MN (2003) Mujer y cárcel en América Latina. In Comisión de Derechos Humanos del Distrito Federal (ed) *Violencia Contra las Mujeres Privadas de Libertad en América Latina*. Washington, DC: Due Process of Law Foundation, 57–74.

Ross R and MacArthur J (2007) *Architecture of Authority*. New York: Aperture.

Salvatore R and Aguirre C (1996) *The Birth of the Penitentiary in Latin America: Essays on Criminology, Prison Reform, and Social Control, 1830–1940*. Austin: University of Texas Press.

Salvatore R, Aguirre C, and Joseph G (2001) *Crime and Punishment in Latin America: Law and Society since Late Colonial Times*. Durham, NC: Duke University Press.

Scott J (1986) Gender: A useful category of historical analysis. *American Historical Review* 91(5): 1067–1070.

Sibley D and van Hoven B (2009) The contamination of personal space: Boundary construction in a prison environment. *Area* 41(2): 198–206.

Stoner J (2012) *Toward a Minor Architecture*. Cambridge, MA: MIT Press.

Tello WM (2010) La cárcel del Buen Pastor en Córdoba: Un territorio de memorias en disputa. *Revista Iberoamericana del instituto Iberoamericano de Berlín* 10(40): 145–165.

Torres L and Herrera M (2010) *Buen Pastor: Una fuga de mujeres*. Córdoba: El Calefón.

Wahidin A (2002) Reconfiguring older bodies in the prison time machine. *Journal of Aging and Identity* 7(3): 177–193.

Wahidin A (2006) Time and the prison experience. *Sociological Research Online* 11(1).

Wahidin A and Tate S (2005) Prison (e)scapes and body tropes: Older women in the prison time machine. *Body and Society* 11(2): 59–79.

Part III

Carceral topographies

The political economy of prison industrial
growth and change

9 Locating penal transportation

Punishment, space, and place
c.1750 to 1900[1]

*Clare Anderson, Carrie M. Crockett,
Christian G. De Vito, Takashi Miyamoto,
Kellie Moss, Katherine Roscoe,
and Minako Sakata*

Introduction

Each penal regime shapes its own spatial configurations, and space also shapes the character of penal regimes. The historical study of this mutual influence opens up for interrogation the "usable past" of carceral geography. For, even as the specific ways in which space and punishment intertwine change over time, their connections remain a fundamental feature of penality in the modern world. This chapter explores these points in a context in which spatiality is perhaps most explicit: convict transportation. Arguably, this penal regime had an even more intimate relationship with spatiality than prisons did, as it bound together convict circulations and geographical contexts through spatial isolation and interconnectedness. Moreover, the routes of convict transportation were often intertwined with other forced labor flows, including African enslavement. The existence of such "scales" of incarceration, migration, and unfree labor were a recurrent feature of transportation across imperial geographies well into the twentieth century (Anderson and Maxwell-Stewart, 2014; De Vito and Lichtenstein, 2013, 2015).

This chapter centers on convict transportation to military fortifications and penal colonies across the British, French, Spanish, Russian, and Japanese empires. The overall time-frame spans *c*.1750 to 1900, a period that witnessed major regime changes, such as the French Revolution (1789) and the Meiji Restoration (1868), and multiple territorial reshaping of the Western empires, especially following the Seven Years' War (1754–1763), American Independence (1776), and the independence of Latin America from Spain between the 1810s and 1830s. Although banishment and exile are practices of punishment that date back centuries, we begin our analysis in the second half of the eighteenth century, which the literature, following Michel Foucault's *Surveiller et Punir* (1975), has long considered the "age of the triumphant prison" (Perrot, 1975: 81; Peters, 2002). We, however, propose that it may be more appropriately characterized as a period of complementary and competing regimes of punishment, and in certain contexts as an age of triumphant convict relocation.

We argue that across various global regions convict transportation may be located within complex webs of punishment, space, and place. First, we investigate how judicial decisions to remove convicts over large geographical distances were connected to ideas and practices of punishment, colonization, and citizenship. Second, we examine why convicts were sent to particular locations, on mainlands, peninsulas, islands, or maritime "hulks," and lodged there in cells, jails, huts, barracks, or forts. We ask which kinds of convicts were sent where, and highlight their spatial mobility as they "progressed" through systems of penal stages. Third, we address how transportation journeys were organized, and the impact that the actual process of transporting convicts over large distances of land or sea had on judicial decision-making. Finally, we consider the question of who made or influenced decisions about convict destinations. We argue that agency in this respect was not restricted to government ministers, magistrates, or judges, but was exercised by other officials and communities, including convicts themselves. Consequently, we point to the need to go beyond a rigid conceptualization of hierarchical "scales" of agency, to argue in favor of a networked and entangled vision of "multi-sited" agencies. Official and proletarian or subaltern experiences and imaginations about the spaces constructed by the networks of transportation are important here. Our broad ambition is to break ground in reconceptualizing transportation as a coherent if locally divergent penal and labor regime underpinned by the dynamics of imperial space across a variety of global contexts.

Sentencing and distance

The idea of removing offenders from their homes and families, and putting them to work in unfamiliar locations, was central to penal transportation as a form of punishment across many centuries and numerous global contexts. Distance was the key feature of this form of punishment, albeit in often contradictory ways. Britain transported convicts to the Americas during the period 1718 to 1775, but following Independence America refused to accept them, leaving Britain to seek out new destinations (Ekirch, 1987). In 1782 Britain experimented briefly and disastrously with transportation to forts in West Africa, and then settled on Australia's Botany Bay (Christopher, 2011). The first convicts were shipped to New South Wales in 1787, with new penal colonies later established in Van Diemen's Land and Western Australia, and transportation continuing until 1868 (Shaw, 1966). Contemporaries expressed the hope that the long voyage to the Antipodes would inspire fear and thus act as a deterrent to crime. Upon finding themselves in a distant and new land, convicts would additionally shed their criminal associations, and, after serving out their sentence, become honest and industrious subjects of Britain's expanding empire.

Comparable views about the importance of distance in judicial sentencing were expressed by the political thinkers G. De Beaumont and A. De Tocqueville in the context of French discussions about penal colonization in the 1830s:

The first requisite of a penal colony is to be separated from the mother country by an immense distance. It is necessary that the prisoner should feel himself thrown into another world; obliged to create a new futurity for himself in the place which he inhabits.

(De Beaumont and De Tocqueville, 1833: 35)

They also acknowledged, however, a key contradiction of distance: that this very separation weakened the "natural ties" between the mother country and the colonies, leading inevitably to the colonies' eventual refusal to receive convicts (De Beaumont and De Tocqueville, 1833: 143). Moreover, as British prison administrator Arthur Griffiths (1894: 4–5) argued at the end of the century, as penal colonies became familiar in the home country, they no longer inspired terror. At the turn of the twentieth century, the Chief Commissioner of the Andamans penal colony, which received convicts from Britain's Indian Empire during the period 1858 to 1939, reiterated this point. He predicted that the Andamans was doomed to fail as the terror of distance subsided, as had been the case in Australia, in another British penal settlement at Singapore (1790–1857), and in the French penal colony New Caledonia, 1863 to 1922 (NAI, 1906).

Notwithstanding the importance of distance, convict destinations were also often determined by the economic and political concerns of the central government. This was the case for Australia and the Andamans, where the British wished to establish a presence for strategic purposes, and did so using convicts. In Spanish America in the second half of the eighteenth century, similar priorities regarding the borders of the empire were met by *presidio* sentences – transportation to military fortifications (Pike, 1983). In the Eurasian sphere, convicts were again used to secure borders in battles over territory. Upon signing the Treaty of Shimoda in February 1855, the Kuril Islands were divided between Japan and Russia. Since the boundary was never marked, Russia settled the northern portion and Japan the southern. In 1868, the first shipment of Russian convicts was sent from the Nerchinsk silver mines to Due Port, which was located on the west coast of the "Russian" region. At the same time, the Tokugawa government suggested to the Hakodate magistrates' office (*Hakodate Bugyo*) that vagrants and criminals be sent to Sakhalin in order to prevent Russian expansion to the south (Tokyo teikoku daigaku, 1922: 429–431). However, the project was never realized, as Japan ceded the island completely to Russia in the treaty of St. Petersburg in 1875. This led the Meiji government to reorientate the country's plans for colonization by convicts to the border island of Hokkaido, as proposed by the Home Minister Toshimichi Okubo (NAJ, 1877). Hokkaido received convicts between 1881 and 1907.

The Russian penal colonization of Sakhalin began in 1862 and ended in 1905, as a consequence of the Russo-Japanese War (Gentes, 2002). Since the reign of Alexander II, legislative framers had theorized that a geographically and culturally isolated prison might be useful in establishing a Russian military and economic presence in the rapidly evolving Far East. The vast actual and imagined geographical and cultural chasm between Russia's eastern-most penal colony and St. Petersburg – 6,500 kilometers and eight time zones – caused both prisoners and administrators to write that they had been exiled to "another world" of an "uncivilized and barbaric" nature (Brower and Lazzarini, 1997: 294). This perception motivated both groups to labor toward "earning" the right to return to European Russia (Gentes, 2011: 303–304). Following his 1890 visit to Sakhalin, Anton Chekhov (1895: 8) wrote, "It is no surprise, that … no one voluntarily travels to the edge of the world [*na krai sveta*]." Sakhalin's inhospitable climate and difficult living conditions functioned as a mechanism whereby understandings of spatiality were both created and deconstructed, and the "natural ties" between the Russian colony and the continental homeland were strengthened rather than weakened.

Destinations

The term "destination," meaning in its simplest sense a place of arrival, may be broadly or specifically defined: as a particular region of empire or as a particular locality. Convicts were sometimes but not always sentenced to a specific destination, and convict destinations could be decided not just by the courts but by administrators, either before or after transportation, on the basis of labor needs. While the networks created by transportation ensured widespread similarities in policies of confinement, the treatment convicts received was inevitably different due to the varied circumstances of different empires, regions, and colonies.

The character of the sentence of transportation from Britain and Ireland changed radically during the period 1787 to 1868, when the last remaining penal colony, Western Australia (est. 1849), closed. What remained constant, however, was the engagement with the use of space as a meaningful component of effective punishment. The system evolved so that convicts were subjected to multi-located stage systems that included cellular confinement and public works in England, followed by transportation to Australia. They progressed through a regime of separate confinement by night and associated labor by day in official hope of moral reformation. Their character was observed in the isolated confines of the transportation ship. They were separated from home, and landed on Australia's shores into relative freedom of mobility and association. And, perhaps most significantly, if they committed further offenses they could be sent to penal stations – islands, peninsulas, or remote locations like Norfolk Island, Port Arthur, or Macquarie Harbour – and confined there in solitary cells or barracks, and worked in immobilizing fetters (Causer, 2011; Maxwell-Stewart, 2008).

Some of the variable penal uses of space may be seen when observing the multiple destinations for those transported to Western Australia from Britain and other parts of the British Empire, alongside the penal transportation of Indigenous people (Aborigines) within Western Australia itself. These destinations appeared in multiple forms – as solitary cells, chain gangs, islands, and hiring depots. The categorization of prisoners – by crime, behavior, or Indigeneity – partially dictated the destinations of these convicts. However, labor needs and policy imperatives were often the deciding factors in the transportation of convicts to specific locales. Each stage of categorization was spatially defined. While incarcerated in Britain awaiting transportation, convicts destined for Western Australia were differentiated by the type of crime committed. For the journey the convicts were segregated once more into divisions to ensure they did not mix (Millet, 2006: 4; Morrell, 1930: 6). Upon arrival in the colony, all convicts were sent to Fremantle Convict Establishment, where they were observed and placed into one of three classes (BPP, 1851: 115). The convicts were further categorized by their conduct (Trinca, 2006). Those classified as "bad" were put in chain-gangs to do hard labor on public works, while those classed as "very bad" were destined for solitary confinement cells. Those deemed "good" or "very good" were rewarded with increased mobility, and they worked as convict warders within the penal colony before being sent out to hiring depots to work for free settlers. These were strategically placed to ensure the even spread of convict labor throughout the colony (TNA, 1857).

Within Western Australia, the containment of the colony's Indigenous convicts began in the early 1830s, when Carnac Island was used as an unofficial site of confinement (BPP, 1837–1838: 191; Moore, 1844: 146). As the frontiers expanded, more and more Indigenous groups were brought up against European conceptions of property rights, resulting in violence and brutal reprisals. The 1840 parliamentary bill that instituted Rottnest Island as an Indigenous prison marked the incorporation of the Indigenous population into British legal frameworks (BPP, 1844: 375–376). As the barrister Edward Landor noted, this hailed the Aborigines as a conquered people subject to British rule, literally clearing the way for the colonial acquisition of territory that was supposedly *terra nullius* (empty land) (1847: 189–192). Unlike the convicts transported to Norfolk, Maria, and Cockatoo Islands for secondary offenses, Rottnest's Indigenous convicts were allowed a high degree of mobility, largely in order to avoid the high death rates which Europeans viewed as the result of their sudden and uncharacteristic immobility (BPP, 1844: 375–376). The security of 18 kilometers of stormy sea allowed an Aboriginal-specific form of imprisonment to arise, as convicts were able to roam and hunt across the breadth of the island, as well as to engage in traditional social activities (such as so-called "corroborees"). A lack of infrastructure for the rapidly expanding colony, rather than ideological concerns about rehabilitation and punishment, resulted in the temporary closure of Rottnest in 1848. Its convicts were transported back to the mainland to be drafted into chain-gangs building the Southern Road to Albany, to construct the new jail in Perth, or to quarry stone for the government school (Green and Moon, 1997: 22). The expense incurred in administering convict islands made their spatial characteristics a generator of continuous conflict between financial and security concerns.

The movement of Australian convicts through penal stages, and the varied use of space in the penal colonies, was part of a globalized set of practices of penal sentencing that worked multi-directionally across and around various parts of the world. Given the regular exchange of information and meetings of international prison experts, it should not surprise us that the sharing of techniques and principles was widespread. From the end of the eighteenth century, for instance, the use of penal stages was adopted in the penal settlements of Bencoolen (1787–1825) and the Straits Settlements (1790–1873), and later on in Burma (1828–1863) (Anderson, 2007). Secondary sites of confinement were established in the Andamans (Viper Island), which received convicts between 1858 and the Second World War, French Guiana (Devil's Island) (1852–1954), and New Caledonia (Île Nou) (Merle, 1995; Redfield, 2000; Sen, 2000; Toth, 2006).

In nineteenth-century Latin America, places of confinement became increasingly specialized (Albacete, 2011; Salvatore and Aguirre, 1996, 2015). Whereas in the eighteenth-century military fortifications (*presidios*) convicts formed only one part of a mixed population of officers, soldiers, Indigenous people (*indios*), missionaries, and free settlers, during the nineteenth century, and especially in urban contexts, the word *presidio* became synonymous with convict-only establishments. Within these institutions, new prison rules sought to foster differentiation between young and adult prisoners, and between men and women. The same holds true for institutions associated with transportation to the frontier zones of post-colonial Latin America (after the 1820s) and in the Spanish Caribbean. Between the end of the nineteenth and the beginning of the twentieth centuries, penal colonies modeled on those of Britain and France emerged in territories where mixed institutions had once been the rule, even when military governors had explicitly called for convict segregation.

In pre-modern Japan, the concept of banishment (*ru/ru-zai/ru-kei*) had been known since the Heian period (794–1185). Corporal punishment and banishment from important cities and trade routes, including the Tokaido and Kisoji roads, were common forms of punishment in the Edo period (1603–1868). Following the Meiji restoration in 1868, a new criminal law called *Shin-ritsu Kouryou* (1870) created punishments of transportation with or without labor for between one and three years. They were abolished under the *Kaitei Ritsu-rei*, a revised criminal law issued in 1873. At this time, apart from capital punishment, imprisonment with or without labor became the uniform punishment for all types of offenders in this revision. However, in further changes to the penal code in 1880, the punishment of transportation with or without labor (*ru-kei* and *zu-kei*) was reintroduced and the practice of sending prisoners to Hokkaido began (Ono, 1880: 7). Spatial differentiation within the Japanese penal colony of Hokkaido (est. 1881) was based on the exploitation of convict labor. Five prisons existed on the island, each with a specific goal: the Kabato central prison was for land clearing and agriculture; Sorachi was for coal-mining; Kushiro was for sulfur mining; Abashiri was linked to road construction; and Tokachi was directed toward agricultural work (Tanaka, 1986: 126–127).

The differentiation between elite prisoners and commoners proved less consistent. In Tokugawa Japan, exile (*onto*) of both elite prisoners and commoners took place in small islands relatively close to the mainland, such as Izu, Goto, Amakusa, Iki, or Oki. Within these locations, exiled people were not confined in special facilities but were free to circulate, albeit under the surveillance of village leaders (*Kumigashira*) (Ishii, 2013: 77–80). By the beginning of the Meiji era Japanese penal reformers aimed to differentiate political and non-political prisoners through transportation to different destinations. During the 1870s, when Hokkaido's establishment was discussed, it was conceived as a penal colony for political prisoners, both as a response to the rebellions of warrior-class people in the Kyushu area (1874–1877) and as a result of the direct influence played by the French model of differentiation of political prisoners (Onoda, 1889: 10–11; NAJ, 1877). During the 1881 to 1886 period however, the Liberty and Democratic Right Movement radicalized, and government policy changed in order to minimize the activists' visibility. Special status was denied to political prisoners, who were sentenced and transported to Hokkaido – along with non-political prisoners – for crimes defined as murder, robbery, and arson (Tezuka, 1982: 128–129).

The lived experiences of different "classes" of prisoners on Russia's Sakhalin reflected a flexibility and freedom that contrasted sharply with the environments of other penal settlements of the same era, even within the Russian empire itself (Gentes, 2002). In contrast to rigid and severe conditions on the Solovki Islands on the White Sea, at the Irkutsk salt works, and in the Nerchinsk silver mines, Sakhalin offered most convicts the opportunity to improve the quality of their penal life (Robson, 2004). Prisoners were classified according to the crimes they had committed, which denoted different living conditions; however, individuals' placements were often assigned with an eye to their potential physical capabilities and skills. Unlike prisoners elsewhere in Russia, all but those who had been sentenced to "life in prison" had the opportunity to ascend the "classification" ladder (pending continued good behavior) and thus retain a measure of personal autonomy. In this way, degrees of penal flexibility uncharacteristic of many prison environments remained within the reach of not only criminal convicts but also political exiles. Exiles who observed the settlement's rules received a monthly living stipend, were generally permitted to move about the island at will, and were free to choose their own housing and employment. Similarly, hard labor convicts could evolve into "convicts in exile" after two years of good behavior. As such, male prisoners earned the right to choose "cohabitants" from among female prisoners, attempt to run small businesses, or seek employment according to their interests and aptitudes. Chekhov observed that numerous prisoners on Sakahlin had become clerks and petty bureaucrats within the Russian prison administration (Chekhov, 1895: 373–374, 395; Doroshevich, 2011: 22, 130–131). The accounts of Doroshevich, Chekhov, and others reveal that some convict settlers established flourishing businesses, traded with the Japanese and Chinese, and even hired employees. When the term of the prisoner's sentence had been fulfilled, he or she became a free settler and could opt to relocate to the mainland, if desired, and settle in Siberia.

Whereas in the case of Sakhalin a common destination featured differentiated penal regimes for political prisoners and hard laborers, in other cases commoners and elite convicts – sentenced for rebellion or political crimes – were transported to entirely separate destinations. Kandyan rebels were removed from Ceylon to Mauritius during the 1810s and 1820s, for example, where they were lodged not in the Indian penal settlement then established on the island, but far away from it (NAM, 1818, 1823; TNA, 1819; IOR, 1813, 1836). Sikhs sent to Burma and Singapore following the Anglo-Sikh Wars of the 1840s were also kept separately from the Indian convicts transported to these locations (Anderson, 2010). During the second half of the nineteenth century, the deposed royal families of India were exiled to Rangoon, Aden, and the Seychelles. Members of the Manipuri royal family, from Assam, who were sent to the Indian penal colony in the Andamans in the late 1800s, were kept away from the convict population (Kothari, 2012: 700–703; NAM, 1880). And yet social status defined the spatial separation of these rebels as much as the nature of their offense; Indian peasant rebels sent to Mauritius and the Straits Settlements during the early nineteenth century were treated in the same way as ordinary criminal transportees.

Similar differentiation influenced the choice of destination, and the very direction of transportation, in colonial and post-colonial Spanish America during the long nineteenth century (AGI, 1764–1780). Elite convicts were rarely shipped from peninsular Spain to the New World's possessions, like the majority of convicts, but figured prominently in the opposite direction. On the eve of Independence in Latin America (1820s–1830s), political prisoners traveled from the colonies to the metropole, although elite political prisoners were usually allowed to remain in peninsular Spain and were not forced to work, while lower class convicts accused of rebellion continued their journey to Ceuta and other minor North African *presidios* and joined forced laborers there. The tendency to expel political prisoners from the mainland continued in post-colonial Latin American states in the form of transportation to overseas penal colonies such as the Galapagos Islands (Ecuador, 1832–1959), the Islas Marías (Mexico, 1905–1939), and Ushuaia (Argentina, 1899–1948), where they sometimes constituted the majority of the convicts (Ortega, 2006; Salvatore and Aguirre, 2015).

We have already discussed the particular spatial confinement of Indigenous convicts in Australia. There were further distinctions of destination for Asian and African transportees. India, for instance, expressly banned the transportation of native convicts to Australia, on the basis that the colonies there were unsuitable for the Indian "race." Europeans born in India were not included in the prohibition, which seems to have been the result of a desire to prevent the influx of tens of thousands of Indian convicts into a predominantly White convict colony (IOR, 1815). The British colonies of the Caribbean sent a few hundred Black convicts to Australia, but in 1837 their presence was declared "injurious" to the majority White population colonies, and further shipments were banned (Anderson, 2012).

Our last point with respect to convict destinations is that penal colonies were often distinctive as homosocial spaces in which women were a small minority, and efforts were made to segregate them. In Australia and the Andamans women were kept in "Female Factories," where they undertook largely domestic forms of labor within a gendered work regime (Reid, 2007; Sen, 1998). Further, across contexts ranging from the convict hulks of Bermuda to the penal colonies of Australia and the Andamans, officials became anxious about the existence of homosexual practices, and introduced new methods of lighting and new routines of watching the closed and often hidden spaces within convict wards. Fears about homosexuality, and, in particular, the so-called moral contamination of juveniles by older prisoners, sometimes led to the transfer of youths to other wards or even destinations. In the Andamans, juveniles were for a period locked up in cages at night in an attempt at segregation. By the 1880s all men and boys labeled "habitual recipients" were confined in separate barracks, and worked separately at stone breaking (Sen, 2000: 173–174; NAI, 1875).

This being the rule, Sakhalin represents an interesting exception. Toward the end of the nineteenth century the Main Prison Administration sought to equalize the 16:1 male-to-female ratio on Sakhalin by sentencing females who had received the milder sentence of exile settlement (Corrado, 2010: 129). Instead of separating the genders in a regulated penal environment, the administration on the island drew them together in bonds of "cohabitation," which, unlike the marriages conducted in penal settlements such as New Caledonia, were neither legalized nor religiously sanctioned. Upon arrival, all female prisoners underwent a selection process during which they were "chosen" by the island's men. Administrators chose new sexual partners from among the women first, followed by certain male prisoners who, because of good behavior, were allowed to choose "cohabitants" with which they would presumably work the land. The only records made of these pairings, however, are single-line entries written in ledgers by administrative clerks.

In this way, Sakhalinian gender politics notwithstanding, female cohabitants often experienced freedoms uncommon to most other colonial contexts. The extra-legality of Sakhalinian marital pairings actually allowed prisoner "wives" the freedom to exert power within – or abandon – undesirable relationships: they often "adopted" orphaned children without spousal approval, carried on multiple, concomitant sexual relationships with men, and made other significant household decisions. Anton Chekhov was astonished to learn, during his 1890 tour of the island, that multiple female prisoners had even murdered their male cohabitants without legal consequences (Chekhov, 1895: 324). He wrote, "nowhere else in Russia is illicit marriage so widely and notoriously prevalent, and nowhere else does it take the peculiar form it does on Sakhalin."

Convict voyages

Travel over large distances of land or sea underpinned the sentence of transportation in all contexts. In Japan, convicts sentenced to exile or hard labor were gathered from all over the archipelago in Tokyo or in the Miyagi central prisons in order to be sent to Hokkaido. The Prison Rules (*Kangokusoku*) issued in 1880 commanded that officials accept convicts three or four times a year, and cautioned against their physical restraint during the journey (NAJ, 1881). Official accounts of life on board the ships are missing, but convict memoirs tell of a man sent to Tokyo central prison in 1887. After being confined there for ten months, he spent a week being transported by ship to Sorachi central prison, alongside 200 others (Koike, 1957: 54–55). Another man was confined in Miyagi for two weeks in 1895 before being transported to Kushiro prison to serve penal servitude for life. In his memoir, he describes his journey from Aomori as follows:

> What depressed me, and what I could never forget, is the scene when on the boat taking us to the ship from the wharf of Aomori…. [We were] looking back to the city as it disappeared from view. We could hear the sound of music and singing from someone's celebration. 260–270 convicts wearing bamboo hats were crouching on the vessel in the rain. Most of them, except a few honest ones who had committed murder, had led a dissipated life and knew the taste of pleasure. They appeared to be on the verge of tears.
>
> (Koyama, 1967: 156)

In 1879 the Russian government began to transport prisoners to Sakhalin by sea. The "Volunteer Fleet," which consisted of seven large steamers of English construction, was assembled in 1878 at the request of Alexander III and began voyages in June 1879 (Corrado, 2010: 72). The steamers' official function was the conveyance of convicts to Sakhalin, as well as colonists, soldiers, and merchandise to other ports in the Russian Far East. The first prisoners to sail, rather than walk, to Sakhalin were a group of 700 who had been collected from prisons "all around Russia" (Corrado, 2010: 71). The press enthusiastically covered the event as the 50- to 60-day trip would be significantly shorter than the two-year walk across Siberia. The ship featured a chapel, a priest, an area for recreation, and a common eating area, although in other ways conditions were deplorable. Vyosovok described the way in which exiles traveled to Sakhalin in cages kept below deck for the duration of the several-month sea journey. In case of mutiny, hoses capable of shooting steam and boiling water were accessible from the main deck (Gentes, 2003: 125). In addition however, the ships were used to carry war munitions and conscripts, causing newspapers such as *The Australian Town and Country Journal* to write that the Russian Volunteer Fleet could potentially be used by the Russian government should hostilities arise in the East, especially since the steamers were being used to transport 16,000 to 17,000 conscripts annually in addition to convicts (1896: 29). Although the Russian government maintained that the steamers were primarily intended to reinforce favorable "commercial relations with the Chinese and the Eastern Siberian ports of the Great Ocean," they were not to be internationally regulated, but remain "freed as far as possible from all useless formalities" (*Nelson Evening Mail*, 1886: 4). As international tensions mounted, the operating conditions of the fleet – despite the Russian wish for non-regulation – were subjected to multinational scrutiny, thus exposing the harsh conditions within which Sakhalin-bound prisoners traveled.

As in the case of the Russian Empire, convict transportation and trade routes were strongly connected in the Spanish Empire, as prisoners were often shipped together with gold, silver, mercury, and fruits, and free "passengers to the Indies" (*pasajeros a Indias*) accompanied by their slaves and domestic servants (*criados*). The "free trade" reform gradually implemented by the Bourbon monarchy during the second half of the eighteenth century greatly impacted upon convict transportation. By authorizing commerce from and to a number of ports in peninsular Spain and Spanish America, and by allowing private merchants to sail between them, it multiplied the routes and nodes of convict circulation and destination. As for trade in general, Cadiz lost its monopoly, albeit not its primacy, in convict transportation, and was joined by Ferrol and Coruña in the northeastern part of the peninsula and by Mediterranean ports like Barcelona and Malaga; the traditional routes connecting Cadiz to the Caribbean (the *Carrera de Indias*) and Acapulco to Manila (the *Galeon de Manila*) ceased to be the only routes of prisoner transportation. Military and non-military convicts were now shipped, for example, from Cadiz to Lima through Cape Horn, and from Galician ports to Buenos Aires. As a consequence, inter-colonial overseas convict transportation increased, especially between New Spain and the Philippines, and in the viceroyalties of Peru and Buenos Aires (where inland convict transportation also played a major role).

The *longue durée* of Spanish transportation also highlights the impact of technological change upon convict transportation, most notably the gradual shift from sailing ships to steamers during the nineteenth century. Besides reducing the time of navigation from 70 days on average in the 1770s to some five weeks in the 1860s along the route Cadiz–Havana, this led to a greater regularity of maritime connections. It minimized the dependence on streams and winds, and therefore on the seasons, and on the limited number of sailing ships available – all factors that had considerably influenced the choices of convict destinations during the eighteenth and the first half of the nineteenth centuries. On the other hand, by progressively lowering the costs of the Atlantic passage, the steamers fostered mass migration of European free migrants to South America at the turn of the twentieth century (Hensel, 2011). Post-colonial governments now channeled the new workforce to borderlands that had been traditional destinations of convicts and other forced laborers; convict labor, meanwhile, was relocated to overseas penal colonies, urban public and municipal works, and within the walls of the new penitentiaries (Salvatore and Aguirre, 2015).

Perhaps the most famous account of transportation from Britain to Australia is that recorded by Dr Colin Browning (1847), who boasted of remarkable success in moral reclamation following his evangelization among the convicts during the long voyage at sea. Australian convict vessels were important sites of religious instruction and industrial training, and, following the appointment of naval surgeons after catastrophic early death rates, for medical intervention and experimentation in disease prevention and hygiene (Foxhall, 2012). It is often claimed that Australian transportation ships experienced remarkably little convict resistance, and yet there is evidence that these mobile yet isolated spaces, designed to confine and to render docile convict bodies, were not always successful. Convicts challenged official efforts to transform them into penal laborers through a daily regime of cleaning, stitching, and oakum picking, by dragging their feet, refusing to work, or erupting in open violence (Maxwell-Stewart, 2013). Officials were always on the watch to counter such resistance, and to ensure the prevention of sexual relations, particularly between men. If ships could be places of moral reclamation, like unlit barracks and shared prison cells, the closed confines of convict berths were also perfect sites of moral contamination (Measor, 1861: 49–50; Tancred, 1857: 6).

Britain's Indian Empire presents a particularly interesting example of the importance of the journey of transportation as a key element of punishment. The British believed that travel over the black waters (*kala pani*, or the ocean) was culturally degrading to "Hindus," since it compromised their caste purity and led to their social death (Committee on Prison Discipline, 1838). In this respect, despite much coalescence in the use of space across global contexts, it is important not to lose sight of elements of local distinctiveness. In the Caribbean colonies which received indentured migrants from India, it is also notable that as a consequence of colonial ideas about the meaning of *kala pani*, transportation was sometimes chosen as a more severe alternative to execution. Officials were of the view that owing to Hindu beliefs about reincarnation, Indians did not express sufficient fear of death on the gallows – but they did fear transportation (BPP, 1876: 225; Kirke, 1898: 225).

For those Aboriginal convicts transported from their homeland to Rottnest Island within the newly demarcated colony of Western Australia, the process of transportation was both drawn out and degrading. The journey was often made largely in chains – whether to the horse of the policeman escorting them, to one another as they walked hundreds of kilometers across the colony, or below the decks of the ships that transported them along the coast (WALC, 1877: 84–85; *The West Australian*, 1887: 3; Commission into treatment of Aboriginal prisoners, 1884: 12–13; Green and Moon, 1997: 48–49). While the state controlled convict movements in the most physical sense on the journey to imprisonment, officials were far less concerned with facilitating their return post-sentence. In 1876, for instance, the legislative council of Western Australia drew attention to the lengthy distances convicts were left to travel once they were released, often through "districts inhabited by hostile tribes" (WALC, 1876: 29). As a consequence of these concerns, the *SS Xantho*, a coastal steamship used for pearl fishing, had a sideline in returning prisoners from Rottnest to the Northwestern territories (Western Australia Museum, 2013). As a symbol of freedom, its importance is clear in its rendering in sandstone at Inthanoona Station that survives to this day.

The geographies of agency

Post-colonial history has centered the question of agency in analyses of empire and imperial societies, exploring subaltern resistance, and tactics of accommodation, as well as the question of whether subaltern voice can be retrieved from the archives of empire (Anderson, 2012; Scott and Tria Kerkvliet, 1986; Spivak, 1993). Post-colonial historical geography has at the same time highlighted the highly mobile character of colonial elites, and their importance in networking empire as they moved across imperial spaces. In this section we bring questions of agency and imperial connectivity to bear on choices of transportation destinations. We argue that multiple, albeit unequally powerful actors, including the convicts and their families and communities, military, legal, and medical officers, and private entrepreneurs, ultimately decided where transportees were sent. We highlight the spatial character of agency, by pointing to the networks of agents that operated contemporaneously across different spaces. This questions predominant visions of agency as located either in imperial centers or in imperial peripheries, divided by rigidly hierarchical forms of institutional power. Different agents clearly had unequal levels of power in the context of the highly segregated eighteenth- and nineteenth-century empires, but the making and implementing of decisions always depended on limited budgets, technological constraints, and officers who variably interpreted both law and their role, and were subjected to contradictory influences "from below." Moreover, convicts, under certain circumstances, were far from powerless but were able to escape, or to use their trans-local social networks to modify imperial decisions through petitioning, and sometimes through collective action or open revolt.

"To the Philippines": this was the sentence of some hundreds of criminal and military convicts transported every year from Cadiz during the second half of the eighteenth century (AGI, 1765–1804; Garcia de los Arcos, 1996; Mehl, 2011). Many reached this archipelago after an average of six months of sailing across two oceans and marching along dangerous inland routes. Once there, the governor general of the islands was officially in charge of establishing their final destination – the military garrison in Manila, the Cavite arsenal, or the *presidios* in the southern island of Mindanao – but jurisdictional conflicts often emerged with local officers and magistrates. Other convicts, however, never reached the Asian archipelago. Some were liberated as a consequence of one of the relatively frequent royal amnesties. Others simply escaped from the San Sebastian castle, public prison, or La Carraca arsenal in Cadiz, or from the royal and private ships that were supposed to transport them. Not all convicts sentenced to transportation sought freedom; some wrote respectful individual or collective petitions begging for a quick transportation and relief from the unbearable conditions of imprisonment.

Some convicts originally destined for Manila were diverted to the Caribbean military fortifications, in the event of wars or after the Puerto Rican hurricane of September 19, 1766 (AGI, 1766). These decisions were made by the King and the secretary of state for the Indies in Madrid, and the *Consejo de Indias* in Cadiz. Much less visible events might result in convicts ending up in the same destinations, or in the North African *presidios*. Imperial officials at all levels often made decisions dependent not on penal factors but on the availability and routes of private merchants, levels of prison overcrowding, the expected arrival of new convicts' convoys (*cuerdas de presidiarios*) from inland Spain, and after medical inspections declared certain prisoners invalided from military and construction work.

Sentencing itself was constructed at the crossroads of multiple influences. As new legal studies on colonial Latin America have shown, legal pluralism and the porosity of the sentencing process itself were such that multiple authorities, prisoners and their relatives, individuals of different status, and sometimes even entire communities were able to influence decisions about the sentence, and even the specific destination of exile and transportation (Benton and Ross, 2013; Cutter, 1995; Haslip-Viera, 1999). One extreme, but ubiquitous, example of this in the Spanish Empire was the case of (elite or non-elite) convicts who were "presented" (*presentados*) to courts by their own parents, most frequently their fathers. They would not only ask the authorities to arrest, imprison, and transport their sons to far-away military garrisons as redemptive punishment for their "bad conduct" and "vice," but would also suggest the exact destination, and even indicate the availability of certain ships in the port for immediate transportation (AGI, 1794). Judicial authorities, for their part, were remarkably keen to meet their demands.

The Spanish Empire is no exception but the norm when it comes to the complex relationship between agency and transportation of eighteenth- and nineteenth-century convicts. Prisoners in nineteenth-century India, for example, displayed a remarkable knowledge of penal settlements, and exercised choice in petitioning for transportation to favored destinations (Anderson, 2005). Some expressed the desire for transfer to settlements with less severe penal regimes, in particular from the notoriously harsh Tenasserim Provinces of Burma to Singapore or Penang (IOR, 1835). It is clear that the Andaman Islands in particular did not become a space of total isolation as intended by the British regime, but a space connected to the mainland through convict letters, and even visits by convicts' families (NAI, 1876).

Agency was not exclusive to convicts and their families. In early nineteenth-century Burma, Mauritius, Singapore, and Penang, officials in the penal settlements competed for the supply of Indian convicts, according to public works requirements (IOR, 1818). Moreover, in some contexts there existed the practice of convict leasing. This was often the result of state convenience as much as the self-promotion of labor contractors. Private capital and governmental priorities, for instance, joined forces in shaping the work-related differentiation in Hokkaido discussed earlier. In 1889 the Horobetsu coal-mine connected to the Sorachi prison was sold to a private company, which continued to use convicts until 1894 (Tanaka, 1986: 113). In the Ishikari Plain, where Kabato prison was located, farms run by private companies or peer cooperatives were created. In 1890, the peer cooperative farm (50,000 hectares) was created in Uryu County, to the north of Kabato prison. Because of labor shortages in the first year, convicts from that penal establishment were leased out in order to construct the related roads (Hatate, 1963: 104–113).

Concluding remarks: penal transportation, spatiality, and the usable past

The concept of the "usable past" is the result of a problem-oriented research methodology, and is not a mere "fact" to be found *out there* in the centuries that have preceded us. In framing and addressing the usable past, we contend, scholars should be aware that "behind every version of the past are a set of inter-ests in the present" – as the editors of this volume put it (2015, 6) – and should refuse any presentist approach to history. In this chapter, we have explicitly avoided any teleological interpretation of historical experiences, and have not made unmediated comparisons of the past with the present. Convicts are still sent over long distances in the present world – in Russia, for instance (Pallot, 2005; Piacentini and Pallot, 2014) – but to argue for a straightforward long-term continuity in practices of penal transportation would be to downplay discontinu-ity in its functions, and in its spatiality. The same point holds for other modes of contemporary deportation and confinement, such as those related to undocu-mented migrants (Gill, 2009).

Here we argue for a more subtle way of framing the relationship between the past and the present, one that in our view allows for a more systematic integra-tion of historical research into carceral geography. We contend that a usable past can emerge: from addressing broad historical and contemporary questions and through appreciating differences in experience across contexts, time, and space. In this chapter we have asked: How do practices and discourses of punishment construct space, and how are the geographies of agency important in the con-struction of punishment? We have argued for a two-fold relationship between transportation and spatiality from a historical perspective. On the one hand, the geographical scope of transportation was broad, making this penal practice a key player in the formation of "networks of empire" (Ward, 2009; Anderson, 2012). On the other hand, precisely because it connected different contexts and involved various types of groups and individuals, transportation had different meanings to different actors across different locations, and convicts played multiple functions in empire building across time and place. The case studies we have presented have revealed these features: they have described convict transportation across oceans and land routes at the same time as they have addressed detailed narrat-ives, individual stories, and specific locations. The integrated study of broad connections and specific contexts is necessary to understand the complex and contradictory spatial experience of penal transportation historically. It is in this methodological "universality" that we find the potential to construct a usable past, while fully acknowledging the irreducibly context- and time-bound nature of the issues that we study.

A similar argument may be put forward regarding the way we conceptualize the geographies of agency in penal regimes. Here we have suggested that a multi-sited, geographically networked, and entangled vision of agency can help scholars address new issues within carceral geography, as it has been in the production of new forms of imperial history in which historical geo-graphers have been so prominent (Lester, 2013; Withers, 2009). This approach invites us to explore the circulation of information upon which officials in London, Calcutta, Madrid, St. Petersburg, and Tokyo based their decisions, and to investigate their perceptions and imaginaries of far-away places that they were unlikely to visit for themselves. Further, it opens out to view the possibility of addressing how convicts made sense of the prospect of transportation and their experiences in the penal colonies, often compared to their knowledge of the nature of punishment in other places, and thus of the connected worlds of imperial, penal regimes. Here again, we find an ideal field for the construction and investigation of a usable past, at the crossroads of various disciplines. Clearly, such large questions, spanning the spaces usually demarcated by particular area studies expertise, can only be answered through collaborative research across the borders of language and empire or nation. For the production of a "usable past" for carceral geography demands recognition of its global history and dimensions. This urges us to take into full consideration the importance of movement across and the opening out of space, as well as its closing in, in the making and experience of punishment.

Note

1 The research leading to these results has received funding from the European Research Council under the European Union's Seventh Framework Programme (FP/2007–2013)/ ERC Grant Agreement 312542. Kellie Moss acknowledges in addition the generous support of the School of History at the University of Leicester in its award of a graduate teaching assistantship to pursue her doctoral project as an affiliated researcher (2013–2017). The following abbreviations are used in reference to archival sources: Archival General de Indias, Seville (AGI); Bengal Judicial Consultations (BJC); British Parliamentary Papers, House of Commons (BPP); Home Port Blair (HPB); India Office Records, London (IOR), National Archives of India (NAI); National Archives of Japan (NAJ); National Archives Mauritius (NAM); The National Archives, London (TNA); Western Australia Legislative Council (WALC).

References

Archival sources

(AGI) Archivo General de Indias, Seville

(1764–1780) Lima, 1524, 1525.
(1765–1804) Arribadas, 287A–B, 548–549, 551.
(1794) Arribadas, 287A.
(1766) Indiferente General, 1907.

(BPP) British Parliamentary Papers

(1837–1838) 685, Western and Southern Australia. Vol. XL. London: William Clowes and Sons.
(1844) 627, Aborigines (Australian Colonies), Vol. XXXIV. London: William Clowes and Sons.
(1851) 1361, 1418, Convict Discipline and Transportation, Further Correspondence. Vol. XLV. London: William Clowes and Sons.
(1876) 1517, Further Papers Relating to the Improvement of Prison Discipline in the Colonies. London: Harrison and Sons.

(IOR) India Office Records, British Library

(1813) F/4/421 10372: Ceylon correspondence, *Board's Collections.*
(1815) F/4/534 12853: Mauritius correspondence, *Board's Collections.*
(1836) F/4/1594 64598: Mauritius correspondence, *Board's Collections.*
(1818) P/133/22, *Bengal Judicial Consultations.*
(1835) P/140/70, *Bengal Judicial Consultations.*

(NAI) National Archives of India

(1875) (1876) (1906), *Home Port Blair Proceedings.*

(NAJ) National Archives of Japan

(1877) *Rutokei o okoshi Hokkaido ni hakken no gi jochin.* Kobunroku 1877, No. 24, Naimusho ukagai 5. February.
(1881) *Kangokusoku kaitei.* Dajoruiten dai 5 hen.

(NAM) National Archives of Mauritius

(1818) RA54: Calcutta letters received.
(1823) RA229: Letters received.
(1880) RA2525: India and foreign correspondence.

(TNA) The National Archives, London

(1819) CO54/73: Ceylon, Governor's Correspondence.
(1857) MPG 1/722: Plans of the Swan River Colony's Convict Establishments.

(WALC) Western Australia Legislative Council

(1877) Parliamentary Debates: Escorting Native Prisoners, 31 July.
(1876) Parliamentary Debates: Rottnest Native Penal Establishment, 9 August.

Secondary sources

Albacete FJB (2011) *La Cuestión Penitenciaria: Del Sexenio a la Restauración, 1868–1913.* Zaragoza: Prensas Universitarias de Zaragoza.

Anderson B (2006) *Imagined Communities: Reflections on the Origin and Spread of Nationalism.* London: Verso.

Anderson C (2005) The politics of convict space: Penal settlements in Southeast Asia. In Bashford A and Strange C (eds) *Isolation: Places and Practices of Exclusion.* London: Routledge, 40–55.

Anderson C (2007) Sepoys, servants and settlers: Convict transportation in the Indian Ocean, 1787–1945. In Dikötter F and Brown I (eds) *Cultures of Confinement: A History of the Prison in Africa, Asia and Latin America.* Ithaca, NY: Cornell University Press, 185–220.

Anderson C (2010) The transportation of Narain Sing: Punishment, honour and identity from the Anglo-Sikh Wars to the Great Revolt. *Modern Asian Studies* 44(5): 1115–1145.

Anderson C (2012) *Subaltern Lives: Biographies of Colonialism in the Indian Ocean World, 1790–1920.* Cambridge: Cambridge University Press.

Anderson C and Maxwell-Stewart H (2013) Convict labour and the western empires, 1415–1954. In Aldrich R and McKenzie K (eds) *The Routledge History of Western Empires.* London: Routledge, 102–117.

Bassin M (1983) The Russian Geographical Society, the "Amur Epoch," and the Great Siberian Expedition 1855–1863. *Annals of the Association of American Geographers* 73(2): 240–256.

Benton L and Ross RJ (eds) (2013) *Legal Pluralism and Empires, 1500–1850.* New York: New York University Press.

Brower DR and Lazzarini EJ (eds) (1997) *Russia's Orient: Imperial Borderlands and Peoples, 1700–1917.* Bloomington: Indiana University Press.

Browning CA (1847) *The Convict Ship, and England's Exiles.* London: Hamilton, Adams and Company.

Causer T (2011) "The worst types of sub-human beings"? The myth and reality of the convicts of the Second Penal Settlement at Norfolk Island, 1825–55. *Islands of History,* Sydney: 8–31.

Chekhov AP (1895) *Ostrov Sakhalin (iz putevykh zametok).* Moscow: Izdanie redaktsii zhurnala "Russkaia mysl."

Christopher E (2011) *A Merciless Place: The Lost Story of Britain's Convict Disaster in Africa.* Oxford: Oxford University Press.

Commission into treatment of Aboriginal prisoners (1884) *Report of a Commission to Inquire into Treatment of Aboriginal Native Prisoners of the Crown in this Colony.* Perth: Richard Pether.

Committee on Prison Discipline (1838) *Report of the Committee on Prison Discipline.* Calcutta: Baptist Mission Press.

Corrado S (2010) *The "End of the Earth": Sakhalin Island in the Russian Imperial Imagination, 1849–1906.* Unpublished doctoral dissertation, University of Illinois.

Cutter CR (1995) *The Legal Culture of Northern New Spain, 1700–1810.* Albuquerque: University of New Mexico Press.

De Beaumont G and De Tocqueville A (1833) *On The Penitentiary System in The United States, and Its Application in France; With an appendix on penal colonies, and also, statistical notes,* trans. F Lieber. Philadelphia, PA: Carey, Lea and Blanchard.

De Vito CG and Lichtenstein A (2013) Writing a global history of convict labour. *International Review of Social History* 58(2): 285–325.

De Vito CG and Lichtenstein A (eds) (2015) *Global Convict Labour.* Leiden: Brill.

Doroshevich V (2011) *Russia's Penal Colony in the Far East,* trans. AA Gentes. London: Anthem Press.

Ekirch R (1987) *Bound for America: The Transportation of Convicts to the Colonies, 1718–1775.* Oxford: Clarendon Press.

Foucault M (1975) *Surveiller et punir: Naissance de la prison.* Paris: Gallimard.

Foxhall K (2012) *Health, Medicine and the Sea: Australian Voyages, c. 1815–1860.* Manchester: Manchester University Press.

García de los Arcos MF (1996) *Forzados y reclutas: Los criollos novohispanos en Asia (1756–1808).* Mexico City: Potrerillos Editores.

Gentes AA (2002) *Roads to Oblivion: Siberian Exile and the Struggle between State and Society in Russia, 1593–1916.* Unpublished doctoral dissertation, Brown University.

Gentes AA (2003) Sakhalin's women: The convergence of sexuality and penology in late Imperial Russia. *Ab Imperio* 2(2):115–137.

Gentes AA (2011) Vagabondage and the Tsarist Siberian exile system: Power and resistance in the penal landscape. *Central Asian Survey* 30(3–4): 407–421.

Gill N (2009) Governmental mobility: The power effects of the movement of detained asylum seekers around Britain's detention estate. *Political Geography* 28(3): 186–196.

Green N and Moon S (1997) *Far From Home.* Nedlands: University of Western Australia Press.

Gribble v. *West Australia*: The Verdict (1887) *The West Australian*, June 30, p. 3.

Griffiths A (1894) *Secrets of the Prison-House or Gaol Studies and Sketches, Vol. I.* London: Chapman and Hall.

Haslip-Viera G (1999) *Crime and Punishment in Late Colonial Mexico City 1692–1810.* Albuquerque: University of New Mexico Press.

Hatate I (1963) *Nihon ni okeru dainojo no seisei to tenkai.* Tokyo: Ochanomizushobo.

Hensel S (2011) Latin American perspectives on migration in the Atlantic world. In Gabaccia DR and Hoerder D (eds) *Connecting Seas and Connected Ocean Rims. Indian, Atlantic, and Pacific Oceans and China Seas Migrations from the 1830s to the 1930s.* Leiden: Brill, 281–301.

Ishii R (2013) *Edo no Keibatsu.* Tokyo: Yoshikawakobunkan.

Kaitei Ritsu-rei, vols I–III (1873) n.p.

Kenney P (2012) "I felt a kind of pleasure in seeing them treat us brutally": The emergence of the political prisoners, 1865–1910. *Comparative Studies in Society and History* 54(4): 863–889.

Kirke H (1898) *Twenty-Five Years in British Guiana.* London: Sampson Low, Marston and Company.

Koike I (1957) Koike isamu jijoden, 2. *Rekishi hyoron* 90: 47–68.

Kothari U (2012) Contesting colonial rule: Politics of exile in the Indian Ocean. *Geoforum* 43(4): 697–706.

Koyama R (1967) *Ikijigoku.* In Kanzaki K. (ed) *Meiji Bungaku Zenshu, Vol. 96.* Tokyo: Chikuma shobo.

Kropotkin PA (1887) *In Russian and in French Prisons.* London: Ward and Downey.

Landor EW (1847) *The Bushman; Or, Life in a New Country.* London: Richard Bentley.

Lester A (2013) Spatial concepts and the historical geographies of British colonialism. In Thompson A (ed) *Writing Imperial Histories.* Manchester, Manchester University Press, 118–142.

Maxwell-Stewart H (2008) *Closing Hell's Gates: The Death of a Convict Station.* Crows Nest: Allen & Unwin.

Maxwell-Stewart H (2013) "Those lads contrived a plan": Attempts at mutiny on Australian bound convict vessels. *International Review of Social History* 58(S21): 177–196.

Measor CP (1861) *The Convict Service: A letter to Sir George Cornewall Lewis, Her Majesty's Principal Secretary of State for the Home Department, &c. on the administration, results, and expense of the present convict system; with suggestions.* London: Robert Hardwicke.

Mehl EM (2011) *The Spanish Empire and the Pacific World: Mexican "Vagrants, Idlers, and Troublemakers" in the Philippines, 1765–1821.* Unpublished doctoral dissertation, University of California-Davis.

Merle I (1995) *Expériences Coloniales: La Nouvelle-Calédonie, 1853–1920.* Paris: Belin.

Millet P (2006) Journeying the punishment, convicts and their punitive journeys to Western Australia 1850–1868. *Studies in Western Australian History* 24: 1–15.

Moore GF (1884) *Diary of Ten Years Eventful Life of an Early Settler in Western Australia*. London: M Walbrook.

Morrell WP (ed) (1930) *Early Days in Western Australia: Being the Letters and Journal of Lieut. H.W. Bunbury, 21st Fusiliers*. Madison: University of Wisconsin Press.

Ono H (ed) (1880) *Dainippon Keihou*. Tokyo: Oshima Saikichi.

Onoda M (1889) *Taisei kangoku mondoroku*. Tokyo: Keishicho.

Ortega AC (2006) *Basalto: Etapa de Terror y Lágrimas Durante la Colonia Penal en Isabela*. Guayaquil: Gráficas Pato.

Pallot J (2005) Russia's penal peripheries: Space, place and penalty in Soviet and post-Soviet Russia. *Transactions of the Institute of British Geographers* 30(1): 98–112.

Perrot M (1975) Délinquence et système pénitentiaire en France au 19e siècle. *Annales ESC* 30(1): 67–91.

Peters R (2002) Egypt and the age of the triumphant prison: Legal punishment in nineteenth century Egypt. *Annales Islamologiques* 36: 253–285.

Piacentini L and Pallot J (2014) In exile imprisonment in Russia. *British Journal of Criminology* 54: 20–37.

Pike R (1983) *Penal Servitude in Early Modern Spain*. Madison: University of Wisconsin Press.

Redfield P (2000) *Space in the Tropics: From Convicts to Rockets in French Guiana*. Berkeley: University of California Press.

Reid K (2007) *Gender, Crime and Empire: Convicts, Settlers and the State in Early Colonial Australia*. Manchester: Manchester University Press.

Robson RR (2004) *Solovki: The Story of Russia Told Through Its Most Remarkable Islands*. New Haven, CT: Yale University Press.

Salvatore RD and Aguirre C (eds) (1996) *The Birth of the Penitentiary in Latin America. Essays on Criminology, Prison Reform, and Social Control, 1830–1940*. Austin: The University of Texas Press.

Salvatore RD and Aguirre C (eds) (2015) Colonies of settlement or places of banishment and torment? Penal colonies and convict labour in Latin America, *c*. 1800–1940. In De Vito CG and Lichtenstein A (eds) *Global Convict Labour*. Leiden: Brill.

Scott JC and Tria Kerkvliet, BJ (eds) (1986) *Everyday Forms of Peasant Resistance in South-East Asia*. London: Frank Cass.

Sen S (1998) Rationing sex: Female convicts in the Andamans. *South Asia* 21(2): 29–59.

Sen S (2000) *Disciplining Punishment: Colonialism and Convict Society in the Andaman Islands*. Oxford: Oxford University Press.

Shaw AGL (1966) *Convicts and Colonies: A Study of Penal Transportation from Great Britain and Ireland to Australia and other Parts of the British Empire*. Melbourne: Melbourne University Press.

Shin-ritsu Kouryou, Vols I–V (1870) n.p.

Spivak GC (1993) Can the subaltern speak? In Williams P and Chrisman L (eds) *Colonial Discourse and Post-Colonial Theory: A Reader*. New York and London: Harvester Wheatsheaf, 66–111.

Tanaka O (1986) *Nihonshihonshugi to Hokkaido*. Sapporo: Hokkaido daigaku tosho kankokai.

Tancred T (1857) *Suggestions on the Treatment and Disposal of Criminals, in a Letter to the Right Hon. Sir George Grey, Home Secretary*. London: T Hatchard.

Tezuka Y (1982) *Jiyuminken saiban no kenkyu; jo*. Tokyo: Keio tsushin.

The Russian Volunteer Fleet (1886) *The Nelson Evening Mail*, April 26, p. 4.

The Russian Volunteer Fleet (1896) *Australia Town and Country Journal*, February 1, p. 29.

Tokyo teikoku daigaku (ed) (1922) *Bakumatsu gaikoku kankei bunsho, Vol. 14*. Tokyo: Tokyo teikoku daigaku bungakubu shiryohensangakari.

Toth SA (2006) *Beyond Papillon: The French Overseas Penal Colonies, 1854–1952*. Lincoln: University of Nebraska Press.

Trinca M (2006) The control and coercion of convicts. *Studies in Western Australian History* 24: 26–36.

Tsukigatachoshi hensan iinkai (ed.) (1885) *Tsukigatachoshi*. Tsukigata: Tsukigatacho.

Ward K (2009) *Networks of Empire: Forced Migration in the Dutch East India Company*. Cambridge: Cambridge University Press.

Western Australia Museum (2013) *Broadhurst Family: An Extraordinary Group*. Available at: museum.wa.gov.au/explore/broadhurst/aboriginal-contact.

Withers CWJ (2009) Place and the "spatial turn" in geography and in history. *Journal of the History of Ideas* 7(4): 637–658.

Wood A (1984) Sex and violence in Siberia: Aspects of the Tsarist exile system. In Massey SJ (ed) *Siberia: Two Historical Perspectives*. London: Great Britain–USSR Association and the School of Slavonic and East European Studies, 22–50.

10 Little Siberia, star of the North

The political economy of prison dreams in the Adirondacks

Jack Norton

Introduction

New York built 39 new state prisons between 1982 and 2000, all of them in rural counties (Department of Corrections, 2014). During those years, the far-northern region of the state, and in particular the 45th State Senate District, staked its future on mass incarceration as a development strategy. More prisons were built in this district than in any other during the prison construction boom of the past 30 years, and the towns and villages of the district competed for prisons with the hope and understanding (encouraged by the Department of Corrections) that prisons would provide economic relief, and growth, to rural counties and small towns that had been hemorrhaging jobs and people at increasing rates since the late 1960s (Huling, 2002). By the turn of the millennium there were 14 prisons located in the district, more than twice the number for any other senate district, rural or urban. And yet the 45th Senate District was not the only area in non-metropolitan New York suffering the effects of deindustrialization, and the communities of the region were certainly not the only ones in the state looking for prisons-as-development in an age of carceral expansion, crime scares, and shifting patterns of production and trade. Why, then, were more prisons built in this district than in any other?

While the lines of the district have been redrawn, slightly, over the decades to account for population shifts – gerrymandered to satisfy political exigencies, including or leaving out this or that town on the periphery – the basic shape of the district has remained the same. It consists of an extensive area north of Albany, bordered on the east by Lake Champlain, and to the north by the Canadian border and the St. Lawrence River. The district encompasses much of the northern and eastern Adirondack Mountains, and is characterized by vast rural spaces, dense forests, and sparsely populated small towns and villages. The economy of the region has been based on, to varying degrees and somewhat in turn, logging, mining, dairy farming, small manufacturing, and tourism. To this list we can add "working for the state," which mostly means working guard duty in state prisons, working for the state police, or working for the Border Patrol.

The 45th Senate District is also far away from the urban areas from which most prisoners in the state originate, which is to say that it is a great distance from New York City *and* Buffalo. The district also hosts the largest and most populous prison in the state, the Clinton Correctional Facility at Dannemora, a town known locally as "Little Siberia" for its climate, remoteness, and for its function within the archipelago of carceral infrastructure in New York (Figure 10.1). Nowadays, after the construction of so many state prisons in the towns surrounding Dannemora, Little Siberia can refer not only to Dannemora but to the entire region, filled as it is with so many prison towns. When Governor Andrew Cuomo announced his first round of prison closures in January 2011, there was an outpouring of resistance, anxiety, and fear from the communities across northern New York, highlighting the extent to which the economy of this northern region is perceived to be tied to mass incarceration (Krieg, 2013; Kaplan, 2011). Why are there so many prisons in this area, and how did prisons become synonymous with progress and development?

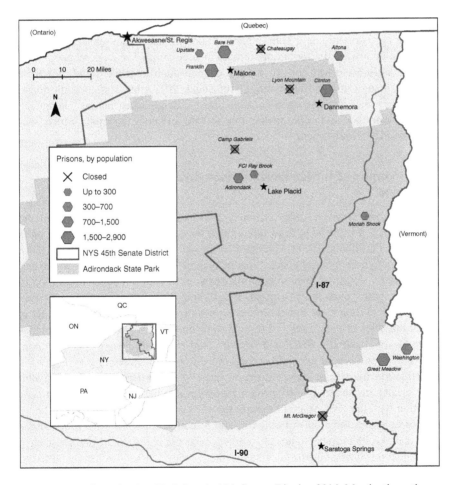

Figure 10.1 Prisons in New York State's 45th Senate District, 2014. Map by the author.

In order to address these questions, it is necessary to look back to the 1980 Winter Olympics in Lake Placid and to highlight the material and political links between the organization of the Games, and the organization of prison expansion in rural New York more generally in the 1980s and 1990s. An understanding of the way the Olympics were brought to Lake Placid helps explain how and why there are so many prisons in the surrounding communities. The history of the 1980 Olympics also shows how a certain model of prison expansion in New York, a model that used prisons as a developmental solution to economic and social crisis in declining rural areas, came to exist as such. This model was by no means inevitable, neither in northern New York nor anywhere else. There is nothing naturally commonsensical in the idea that a prison – an institution that cages human beings and removes them from civil life and from the labor force – would be an engine of enhanced social capacity, development, and freedom. The building of a federal prison in conjunction with the 1980 Winter Olympics is thus a story of how prisons became synonymous with development in New York, and how the construction of this new common sense drew on already-existing infrastructure, and both expanded and shifted already-existing political organization toward the formation of mass incarceration. The carceral geography of New York State was constructed not only through intensified and racialized urban policing (Kohler-Hausmann, 2010; Segal, 2012), but also through northern communities drawing upon a usable past of social infrastructure – hospitals, schools, mines – and repurposing this toward prison expansion, which was understood as development.

The Olympics of Little Siberia: prison dreams and carceral expansion

To understand how and why so many prisons were built in and around the Adirondacks, it is useful to draw attention to the ways in which certain structures of feeling (Williams, 1977) were created through prison expansion in northern New York. These emergent structures of feeling – understandings of limits and possibilities – in turn helped to naturalize prisons as a common-sense solution to problems both "over there" (urban, Black, and Latino New York) and "over here" (rural, White New York). Prison expansion in New York relied heavily on racial fantasies of purity and uplift, as well as on rural, White understandings of crime as an essentially Black and Brown – and above all *urban* – phenomenon. These themes, brought up again and again by communities seeking to justify the construction of state prison facilities in the 1980s and 1990s, were articulated very clearly in the struggle to bring a federal prison to the Adirondacks in conjunction with the 1980 Winter Olympics.

For example, the plan to build a prison in conjunction with the 1980 Olympics in Lake Placid was opposed by many who made the claim that the nearly all-White and rural population of the northern Adirondacks was poorly suited to "guard" a prison population of disproportionately Black and Brown people, and that the prison would turn these same White northern New Yorkers into "the herdsmen of minorities." In response to these criticisms, both the Saranac Lake and the Lake Placid Clergy associations signed a statement in support of the Olympic Prison, claiming that not only were residents of northern New York not racist, but that they would draw upon a tradition of "serving and helping people" in their work as guards (Local Clergy Backs Prison, 1978). These caring skills, according to the clergy, were in particular demand given the then-current over-crowded and "inhumane" prison conditions, a problem whose solution was to be found in the Olympic prison. According to these local religious leaders, that the prison needed to be built at all resulted not from structural forces but from "the harsh realities of human evil." However, that the prison should be built near *Lake Placid*, according to these same leaders, was because it would help address "chronic unemployment" in a region for which "the Government [had] done too little" (*Lake Placid News*, 1978).

The story of how the realization of the dream of the Olympics at Lake Placid was made possible through prison construction is an important moment in the emergence of mass incarceration as a developmental solution to the problems that faced northern New York. Prisons are carceral infrastructures, and mass incarceration consists of various physical, political, and judicial infrastructures that come together to produce a system that cages people to temporarily and spa-tially resolve crises at different scales. Focusing on infrastructure aids us in con-sidering what else *could have been done* with this social capacity for organization and for development that was ultimately channeled into the produc-tion of mass incarceration as a "catch-all solution to social and economic prob-lems" (Gilmore, 2007: 5).

The political and material practices that brought the 1980 Winter Olympics to Lake Placid were transitioned to a project of carceral expansion in the years fol-lowing the games. Indeed, prison expansion in northern New York was an integral part of the games themselves, as the organizers of the Winter Olympics in Lake Placid arranged for the athletes' housing to be constructed so that it would be used as a federal prison immediately following the Olympic competition. Numerous state prisons were built in close proximity to the Olympic Village in the years fol-lowing, often through reuse of the abandoned infrastructure of public schools and hospitals. Unlike many communities in other states where local resistance often accompanied prison siting, northern New York towns and villages actively organ-ized to attract prison "investment." For example, Franklin County, a rural county in the 45th district, to the north of Lake Placid, saw four new prisons built within its borders between 1986 and 1999 – three in the Town of Malone, and one in the nearby Town of Chateaugay. These prisons, which transformed Malone and Cha-teaugay into prison towns, were built after intensive lobbying on the part of local small business owners and politicians who saw these prisons as an opportunity for local job creation. While just one piece in a larger history of mass incarceration in New York, an understanding of how these local (and rural) growth machines formed and re-formed the carceral landscape sheds light on the ways in which common-sense understandings of race and development are created and mobilized toward certain material ends (Gramsci, 1971).

Scholars studying mass incarceration in the United States have pointed to the prison expansion of the past 40 years as a state-led developmentalist solution to social and economic crises in an era of a neoliberal rolling back of social services and rolling out of increased policing (Bonds, 2006; Gilmore, 1998; Morrell, 2012b; Wacquant, 2009). Prison expansion in rural areas has gone hand in glove with disinvestment in social spending, even as prison siting in rural communities was understood as a way to develop local economies and create jobs (Hooks *et al.*, 2010; Lawson *et al.*, 2008). New prison construction was often "sold" to rural and deindustrialized communities as a way to encourage local development, as a way to attract "recession-proof" jobs funded by the state in an age of ever-widening criminalization (Huling, 2002; Bonds, 2009; Gilmore, 2007). The evidence suggests, however, that prison siting in rural areas actually slows down local economic growth even as it reworks the relationship between the urban and rural in changing patterns of state spending and disinvestment (Hooks *et al.*, 2004; Gilmore, 1998; King *et al.*, 2003; Besser and Hanson, 2004). Less evident in the unemployment statistics, however, is the experience of being in a "prison town," on the outside of the walls, a feeling that Fraser, writing about Susanville, California, described as "a pervasive sense of antagonism" (Fraser, 2000). While prisons do provide some jobs in northern New York in the midst of still-ongoing outmigration and economic decline (which may have been spurred by the prisons themselves), the main streets are often boarded up, and the towns eerie shells of themselves. Malone, NY, a town where three state prisons were built between 1986 and 1999, recently identified the outmigration of young people as one the biggest problems posed to the sustainability of the community. Yet this was exactly the problem that prisons were supposed to solve in this northern region (Camoin Associates, 2009).

Gilmore (1998, 2002, 2007) explains the buildup in California – the largest expansion of carceral infrastructure in history – as a spatial "fix" to economic and social crises in an age of "post-Keynesian militarism." Gilmore demonstrates how prison construction in rural areas of the state temporarily resolved surpluses in rural land, urban labor, financial capital, and state capacity. These four surpluses were "fixed" – both in the sense of physically fixing them to a place, and in the sense of fixing the problem of lost value that would result from the inability of capital to dis-accumulate – by constructing prisons on an unprecedented scale (Gilmore, 2007; Harvey, 2007). As fixed capital in rural landscapes and as places made for and made by "human sacrifice," prisons become a geographic solution to crises of overaccumulation – uneven crises whose resolutions are always contingent and determined through struggle (Harvey, 2006b; Gilmore, 2007).

The story of the Olympics in New York is a story of economic development, of people organized to attract capital – in the form of state investment in carceral infrastructure – in a time of crisis. This is also a story about mass incarceration in New York, about a prison construction boom that saw the prisoner population in the state increase nearly six-fold. Finally, it is a story of the re-valorization of fixed capital in a rural landscape that was in the process of being forgotten and abandoned (Harvey, 1996). This historical geography of prisons in northern New York is written from a historical materialist view of the built environment and cultural change. Our present conjuncture rests upon the outcome of past struggles, and in this sense we still struggle with the future of the usable past.

Lake Placid, 1974: Olympic dilemma, prison fix

Lake Placid is a village of just under 3,000 residents, located in the middle of the Adirondack Mountains in northern New York. Home to the 1936 and the 1980 Winter Olympics, the village lies within the borders of rural Essex County, and is in the 45th Senate District of New York State. To reach Lake Placid by car one must drive through some of the highest mountains in the Adirondacks on a two-lane road. The closest village to Lake Placid is Saranac Lake, about six miles away. In between Saranac Lake and Lake Placid is the hamlet of Ray Brook, and at Ray Brook, hidden from view, are two prisons – one federal and one state – as well as the headquarters of the Adirondack Park Agency. That Lake Placid is a prison town is difficult to discern if you are not there for the prison. Most visitors to the village come for the mountain views and for the Olympic facilities – including a ski jump, bobsled run, and the ski slopes. However, the Olympic and carceral landscapes of Lake Placid share not only geography but a common architecture and history. The juxtaposition of the Olympic facilities and prisons is not coincidental. The leadership of Lake Placid, along with the Bureau of Justice and state officials, built a prison in conjunction with the games, as a solution to logistical problems posed by the games themselves. The prison, like the Olympics, was seen as part of a solution to a local crisis of social reproduction, brought on by northern New York's shifting relative location in global circuits of trade and production. The crisis was experienced and described, locally, in terms of high unemployment rates, deindustrialization, and outmigration. The building of the federal prison at Ray Brook also provided communities across northern New York with a model of how to deal with such crises, through the construction of state prisons and employment of guards. Many of these small mining and agricultural towns transitioned to prison towns in the years following the games.

   Lake Placid is a small community, and local community leaders – small business owners, local politicians, and civic-minded residents – had attempted to attract the Olympics to the village before 1980. They had bid for the 1960 Games in 1954, but the Olympic Committee chose Squaw Valley, California. They tried for the 1968 Games, but they went to Grenoble. They attempted to convince the United States Olympic Committee to make a bid for Lake Placid to host the 1976 Games, but the Committee chose Denver instead. When Denver backed out, the National Committee chose Salt Lake City as a replacement, but the 1976 Games were played in Innsbruck, Austria. By the time Lake Placid won the bid to host the 1980 Games – and they made and won that bid in 1974 – the village's Olympic Games Committee had attempted to lure the games to their small village seven times in 20 years (Johnson, 1974; Stafford Papers, S2, B1, F19, 1977).

The Olympic Task Force that bid for the 1980 Games was thus well versed in the ins and outs of Olympic politics. In a 1974 article in *Sports Illustrated*, these "North Country Boys" were portrayed as a group of determined small-town men who tenaciously evangelized and organized for a Lake Placid Olympics, despite being a rather scrappy group from a town that, without the Olympics, would otherwise be just another "frozen flyspeck on the U.S. map" (Johnson, 1974). That the 1980 Olympics took place in Lake Placid was remarkable given the town's size, remote location, and lack of political connectedness. At the same time, the Olympics allowed for the strengthening and expanding of local political machines, and it served as a national and even international spotlight for local politicians. Money flowed into the area in the years leading up to 1980, as the Games ended up costing US$168 million, well over the initial estimate of US$30 million (Gish, 2006). In addition to the money, the national and international attention the games attracted served as an opportunity for talented local politicians to expand their influence in the state. Not only was this influence used to build a prison in conjunction with the Olympic Games, but it was also used to direct prison expansion to northern New York when the opportunity presented itself in the following years. One of the most well-placed, talented, and charismatic figures in these years was State Senator Ronald Stafford, a man whose origins in Dannemora and career in the 45th Senate District parallel the transition that the northern Adirondacks made from the Olympic Games to prison expansion.

Stafford represented the 45th Senate District from 1965 to 2002, becoming an extremely powerful figure in the State Senate by virtue of his seniority and his considerable political talent (Dao, 1993). He was a Republican Deputy Majority Leader for a time, and was Chairman of the Senate Finance Committee for almost a decade. He was responsible for an enormous amount of legislation, including New York's Tuition Assistance Program (TAP grants), and he ran numerous "tough-on-crime" campaigns throughout his career (Stafford Papers, S3, B3, F2, 1985). Stafford passed away in 2005, but the memory of the powerful and personable senator looms large in the northern Adirondacks. In discussions of the recent prison closures in New York state, many local politicians and community leaders claim that "if Ron Stafford were alive, [these prison closures] wouldn't be happening" (Anonymous A, 2013; Anonymous B, 2014). Stafford brought in hundreds of millions of dollars of development money to his beleaguered, deindustrializing district during his four decades of public service, and in so doing created a powerful and entrenched local political machine. Stafford also used his influence to turn the northern Adirondacks into a penal colony for the sake of job creation, a region to which prisoners' families must ride buses from New York City for up to ten hours at a time to catch glimpses of their loved ones. Where Stafford really began to assemble the pieces of this political machinery was in bringing the Olympics to Lake Placid.

Stafford was the son of a prison guard, and he grew up in Dannemora, where the houses on one side of the main street face the prison wall on the other side. The prison at Dannemora, Clinton Correctional Facility, was built in 1844. As noted above, Dannemora is the original prison town in northern New York; that is to say, it is a town where the prison provides the majority of the jobs and is the institution that ultimately structures the life of the community outside of its walls. The town is located within the Adirondack Park boundary line. The region was once a center of iron mining, when iron ore was shipped by rail and canal barge to foundries downstate. The small neighboring village to the west of Dannemora, Lyon Mountain, had been a company mining town until Republic Steel closed down in 1967, leaving the abandoned gray mine buildings looming on a hill above rows of company-constructed houses. The prison at Dannemora had in fact been built there in order to facilitate the use of convict labor in the nearby mines. This practice, unprofitable and contested by organized labor (which objected to and organized against competition from bonded labor), was discontinued in 1877 (Glynn, 1995; Lewis, 1965). There is no mine at Dannemora now, only a maximum-security prison, and where mines once existed at nearby Lyon Mountain and Moriah (towns both in the 45th District), prisons were built in the 1980s. Where once prisoners left the prison to work in the mines, now the sons and daughters of miners go into the prisons to work as guards (and are also themselves prisoners).

Stafford was first elected to the State Senate in 1965 (Thompson, 2005). In the early 1970s, he began organizing to bring the Olympics to the Lake Placid region, and he was made Chairman of the New York State 1980 Olympic Winter Games Committee in 1974. Stafford put together and led a provisional Olympic task force, which then prepared to present their case at the International Olympic Committee in Austria. The task force comprised local civic leaders and small-business owners. Stafford, the most well-connected member of the Committee in terms of state politics, was still in the early stages of his political career, and did not yet have the political power that characterized his later incumbency. Many of the task force members had participated in the 1936 Lake Placid Olympic Games as young athletes and were eager to bring the excitement and activity of the Games back to the area (Johnson, 1974). These "North Country Boys," mainly local petit-bourgeoisie and officials, came together to amplify and project their power for the sake of both the Olympic ideal and economic growth (Stafford Papers, S2, B1, F19, 1977).

Lake Placid's bid for the games was a long shot; it was a remote, isolated little town in the middle of the Adirondacks with a population of under 3,000, making it one of the larger villages in the area but an extremely small venue for such an event. The village had hosted the Games in 1936, when the Games had been smaller and less visible (untelevised), and it still served as a regional training center for winter athletes. Since 1936 though, the Olympics had become more of the highly capitalized spectacle that they are today – events backed by large public debts which can weigh down host communities for years; ephemeral gatherings made possible by spending on massive and costly infrastructure; projects which often displace local people and which are just as often underused once the Games have finished. These were some of the primary reasons that voters in Colorado had rejected the Games in 1972, and members of the

Committee in Lake Placid were also concerned that the Olympics might saddle the area with debts that would long outlast any benefits derived from the few weeks of activity. It was in this context that the Ron Stafford's Olympic Task Force, located as it was in a "tiny mountain village," decided to pitch the Lake Placid Olympics as a return to a purer, more athlete-centered competition, in contrast to the huge spectator-driven, debt-fueled Olympics of the recent past (Stafford Papers, S2, B1, F34, 1975).

The Task Force also saw in the Olympics a solution to an economic crisis; the Olympics could serve as a tool with which to address the very high levels of unemployment in the region. By the early 1970s, those northern counties were in the midst of a wave of deindustrialization and were suffering the effects of mine and factory closures. The unemployment rate in the area in the early 1970s was 16 to 17 percent – 19.9 percent in Hamilton County – compared to a state average of 9 percent. The Committee used these economic arguments for the Olympics at Lake Placid to make their case to New York State and then to the National Olympic Committee, claiming that preparations for the Olympics would bring "275,000 man-days of work in construction alone" to the area, providing, in a four-and-a-half-year period, "the equivalent of 250 full-time jobs." The committee argued that "for the period six months preceding the actual events, 425 new jobs" would "come into existence" (Stafford Papers, S2, B1, F34, 1975).

When the decision to award the 1980 Olympics to Lake Placid was made in 1974, the Olympic Task Force continued on as the organizing committee, with retired postmaster Ronald Mackenzie as President and Senator Ronald Stafford as a member of the Executive Committee. This committee was soon faced with certain key challenges to organizing the games, namely the need to plan, coordinate, and build crucial infrastructure in such a small village. While much of what was needed for the winter Games already existed in Lake Placid – the ski slopes, a bobsled run – the Committee needed to modernize many of the facilities, and to construct a ski jump. In addition to the sports infrastructure, the Committee was faced with the problem of how to house all the people who would be coming to the area for the Games. The population of the small village was expected to briefly double to 6,000 during the Games. Some of this demand could be met through existing hospitality infrastructure, and a committee of motel and restaurant owners was formed to coordinate lodging, not only in Lake Placid but across the region. The Adirondacks are vast, and housing was also needed for the participating athletes. It was especially important that these athletes' housing be close to the Games, in Lake Placid, and not a long commute away in nearby towns over what would be traffic-clogged two-lane mountain roads.

The Committee's concern was – and had been from the initial meetings – to avoid spending millions of dollars in public money to build infrastructure that would be abandoned afterwards. This concern stemmed from the fact that they had made the case for a Lake Placid Olympics arguing against just this sort of spending, and with recent mine and mill closures the Adirondacks and surrounding region was already littered with the remains of shifting capital flows: empty mines, decaying infrastructure, and vacant housing. Ongoing deindustrialization meant a proliferation of empty buildings in the rural landscape. In addition, the main impetus for the Olympic bid, as articulated by the initial Task Force Committee and supported by the population of Lake Placid, was to develop the area in a way that would benefit the local economy in the longer term. Development in the area, at a time when unemployment was twice the state average, meant jobs that would allow people to continue living in the small rural communities in and around the Adirondack Mountains.

At an organizing committee meeting in January 1976, Ronald Mackenzie presented a possible solution to the Olympic athletes' housing dilemma. The Ray Brook Sanitarium near Saranac Lake was slated for closure, just four miles away from Lake Placid. Mackenzie, concerned about both where to house the athletes and the looming specter of an abandoned hospital on the outskirts of town, suggested that they invest money set aside for the athletes' housing in order to modernize and clean up the Ray Brook sanitarium for that purpose. Makenzie suggested that the sanitarium be used to house the visiting athletes with a view toward re-purposing the structure as a community college post-Olympics. North Country Community College, located in Saranac Lake, was dispersed in various miscellaneous buildings around town. By consolidating and expanding the college into a renovated Ray Brook Sanitarium, the problem of the athletes' housing would be solved in a way that would prevent blight, and in a way that would provide long-term employment to residents of Saranac Lake and Lake Placid (Stafford Papers, S2, B1, F38, 1976). Mackenzie noted, however, that the buildings at Ray Brook were more spacious than necessary for the community college, even though the sanitarium was the perfect building, in the best location, for the Olympic Athletes' Village.

The sanitarium at Ray Brook did end up serving as housing for visiting athletes during the 1980 Olympics, and it was used as a federal prison beginning immediately after the Games and remains so today. Sometime early in 1976, after talking with various federal agencies about possible after-uses of the Olympic housing site, the Olympic Organizing Committee began talks with the Federal Bureau of Justice to explore how the Ray Brook sanitarium might be re-purposed, first as an Olympic Village, and then as a federal prison post-games (Stafford Papers, S2, B1, F32, 1976). The prison at Ray Brook was officially proposed at a meeting on June 1, and Norman Carlson, Director of the Federal Bureau of Prisons, toured the site the following day. The deal was finalized by the autumn of 1976 (Stafford Papers, S2, B1, F19, 1977). The Department of Commerce contributed US$12 million to improve Ray Brook to the standards of Olympic athletic housing, and the Department of Justice contributed an additional US$7 million to renovate it to the specifications appropriate for an after-use as a federal medium-security prison (Stafford Papers, S2, B1, F29, 1976).

Assembling this complicated and time-sensitive deal required a level of political connection and brokerage that the members of the Organizing Committee, with the exception of Ronald Stafford, did not possess. It is here, during the lead-up and planning stages of the 1980 Olympics, that Stafford's profile and influence expanded. In October 1976, Mario Cuomo, the future Governor and, in

1976, New York's Secretary of State, assisted the Olympic Organizing Committee by making a case for the federal prison after-use on the grounds of jobs and economic development. According to Cuomo, the federal prison at Ray Brook would "provide much needed benefit to the Essex County area as well as further the important development of the 1980 Winter Olympics" (Stafford Papers, S2, B1, F32, 1976). Cuomo and the Task Force's framing of the Olympic prison as an economic development issue – and as a move to eliminate blight – would turn out to be significant in subsequent years as the state struggled to finance prison expansion.

The Olympic model: prisons-as-development in the Empire State

With the conversion of the Ray Brook Sanitarium into the Ray Brook Federal Prison, the Olympic Task Force, led by Stafford, and with the help of Mario Cuomo, solved the problem of the athletes' housing, while at the same time addressing the issue of abandoned infrastructure in the northern landscape. Indeed, the "North Country Boys" of the Olympic Organizing Committee were able to utilize their new, and ephemeral, position at the center of attention in order to leverage federal funds to re-valorize about-to-be-abandoned fixed capital in the Adirondack Mountains. They also made a gesture at addressing unemployment in the region by adding "216 jobs." Through this alchemy of the sanitarium's conversion to athletes' housing with a carceral after-use, the meaning and goal of the Olympics and incarceration converged as economic and infrastructural development for the region.

The Olympic Prison at Ray Brook served as a model for prisons-as-solution to both unemployment and the abandonment of rural areas by capital. Carceral infrastructure provided a technical solution to political and social problems in both rural and urban areas of the state, even if these places seemed disconnected. The appearance of their disconnection was, and still is, fundamental to the ways in which prison expansion was made to appear as a "common-sense" solution to urban crime and rural unemployment. The Olympics-to-prison model showed how fixed capital could be re-valorized in the North Country through these new political alignments between rural growth machines – local boosterism like that of the "hick town" North Country Boys (Johnson, 1974) – and liberal Democrats looking to expand carceral infrastructure and capacity through public–private partnerships (public debt issued through public authorities, without voter controls) (Schlosser, 1998; Morrell, 2012a).

Locally, the 1980 Olympics also provided Stafford with the means – money, jobs, and influence – to build an effective political network. This network relied on cooperation with Governor Hugh Carey and Secretary of State Mario Cuomo, but was also constructed on a foundation of already-existing relationships. This growth machine – an alliance of local politicians and the rural petit-bourgeoisie – was later on to be further mobilized toward mass incarceration as a solution to economic decline. It is a political formation that appears now in northern New

York as part of a naturalized order of power, and is often referred to as "the powers that be" by local residents. The presence of Clinton Correctional Facility at Dannemora was also central to this emerging local common sense of prisons-as-development in the northern Adirondacks, since existing political networks already included men who had lived within the shadow – and many who had worked within the walls – of the largest maximum-security prison in the State of New York. In the wake of the decision awarding the 1980 Olympic Games to Lake Placid, Stafford received many letters requesting positions in the various Olympics-related jobs that were becoming available. For example, the man Stafford recommended for Director of Contract Services, for his administrative abilities, was Ed LaValley, who had worked for New York State Corrections for 50 years and who "ran Dannemora," the prison in Stafford's home town, "without any problems" (Stafford Papers, S2, B1, F32, 1976).

As prison expansion proceeded apace in New York State in the early 1980s, more prisons were built in and around the northern Adirondacks. At the same time, the process of prison expansion began to resemble the Olympics-to-prison "program" in Lake Placid in terms of both how and why new prisons were built in this area. Abandoned rural infrastructure – usually schools, hospitals, or mining buildings – was converted into carceral infrastructure. In Stafford's 45th Senate District, the Camp Gabriels tuberculosis sanitarium, in the southern section of Franklin County, was converted into a prison camp in 1982. Altona, New York, a small village in the northeast of Stafford's district, near the border with Quebec, was losing population through out-migration. Altona's central school was closed and then re-purposed as a medium-security state prison in 1983. Lyon Mountain, the company mining town next to Dannemora, continued to lose population in the wake of the mine closing, and its central school was converted into a minimum security prison in 1984. The Department of Corrections used the idled mining structures of the village of Moriah, whose mine had closed in the early 1960s, to create the Moriah Shock prison in 1989.

Conclusions: infrastructural development and limits to state capacity

The power of New York to expand the state prison system in the early 1980s was limited by the electorate's lack of enthusiasm for public debt. The prison population in New York State climbed steadily in the wake of increased policing in urban communities as well as the mandatory minimum sentencing laws that had been put in place in the mid-1970s by Governor Nelson Rockefeller. As the prison population rose, the Department of Corrections attempted to increase its capacity to keep people locked up. While many argued that the prison over-crowding problem could be solved through reforming the parole system and rolling back mandatory minimum sentencing (Correctional Association of New York, 1982; McDonald, 1980), the New York Department of Corrections presented Governor Carey with a prison expansion plan in 1980; technical solutions to political problems. In order to fund this prison expansion program, Carey put a US$500 million general obligation bond issue to a state-wide referendum, which was required by the state constitution. The Correctional Association of New York led a campaign against the bond issue, and the state prison guards' union – the New York State Correctional Officers and Police Benevolent Association (NYSCOPBA) – as well as the Department of Corrections itself, also campaigned vigorously in favor (Jacobs and Berkowitz, 1983). The bond issue was defeated in 1981, and the state's capacity to expand the prison system was curtailed.

Mario Cuomo found a way through this limit on state capacity, and he did so, just as he had done for Lake Placid, by arguing for prisons as a solution to unemployment and deindustrialization. When Cuomo became Governor of New York in January 1983, he was immediately confronted with a crisis in the form of a prison rebellion at the Ossining Correctional Facility (Sing Sing) that lasted four days. In the various inquiries and reports that came out in the wake of this "disturbance" at Ossining, prison overcrowding was identified as the main cause. Cuomo responded by announcing his intention to utilize the Urban Development Corporation (UDC) to finance a massive prison construction boom in the state. The UDC is a powerful public authority created in 1968 by Rockefeller to overcome voter resistance to funding low-income housing in urban areas, and it has the ability to circumvent the New York State Constitution by issuing bonds without approval through a state-wide referendum. Cuomo argued that since prisons would bring jobs to deindustrialized areas in upstate New York with high unemployment rates, financing prison construction fell under the mandate of the public authority to develop blighted areas. He was granted approval by the State Legislature, and every prison built in the state between 1983 and 2000 was financed through bonds issued, without voter approval, by the Urban Development Corporation.

There was never much of a problem finding places to locate prisons in New York, because there was local, "grassroots" support for prisons-as-development in many communities in the state (Millard, 1986). This was especially true in the towns and villages of Ronald Stafford's district (Schlosser, 1998). In fact, when one looks at how communities in Stafford's district actually competed with one another for prison siting, it is clear that they had the same form of organization – prison "task forces" that utilized local letters of support, including from schoolchildren – and which were mobilizations of the petit-bourgeoisie in those areas. These prison task forces had the same class composition and the same form as the original Olympic task force from which Stafford built his political organization, and they were used by Stafford as a way of making a case in Albany for more prisons in his district while also making a case to constituents that he was providing development money and bringing jobs to northern New York.

Through the mobilization of this political machinery, communities in northern New York facilitated the expansion of a system designed to immobilize labor, and they did this by trying to capture increasingly mobile capital and commodity flows. Prison expansion in rural New York was thus supposed to open up possibilities of development and freedom for some people and places, to the detriment of others. In 1985, when construction on the first prison began in Malone, a town in the 45th District that is now home to three prisons, a series of editorials in the local paper proclaimed that Malone was "on the move" (Malone Farm Sold, 1989; Boyer, 1986).

From the vantage of Lake Placid in the 1970s, the Olympics and prison construction converge at economic development, and both end up being schemes to re-valorize fixed capital, to provide jobs, and to redirect value (through the state) onto the landscape. While always a small mountain village, Lake Placid's relative location – along with northern New York in general – was shifting along with broader networks of production, consumption, and transportation at the end of the so-called Fordist period of American capitalism (Harvey, 2006a, 2007; Gilmore, 1998). Like the infrastructure of the 1980 Olympics – of which the Ray Brook Federal Prison was part – prisons in the North Country have served as places to dis-accumulate surplus capital as well as a way to reshape the flows of labor and capital at a regional level.

Taking a longer view of infrastructural development in New York State, the prison boom of the 1980s and 1990s takes place within a longer history of public works under capitalist development, such as New York's canal system (McGreevy, 2009), waterworks (Gandy, 2002), and parkway and highway network (Caro, 1974). Canals and highways are built in order to keep capital moving in the commodity form, and to speed up circulation so that surpluses do not over-accumulate. As Marx explained in *Grundrisse* (1993), highways and canals are state infrastructures that allow such movement and facilitate the "annihilation of space by time." Prisons, however, are an infrastructure, also directly facilitated by the capitalist state, that annihilates time by space, and that fixes surpluses of labor and capital in place. And they also fix, as Gilmore points out in California (Gilmore, 2007), surplus land, or surplus infrastructure. In the case of New York, this surplus land and infrastructure often already appeared as public or social infrastructure. Thus the re-purposing of Olympic housing, of hospitals, and of public schools in the North Country, in yet another way and in another place, highlights just how wasteful this system of mass incarceration has been in terms of human life and social capacity for development.

Coda: upstate/downstate

Anyone from upstate or western New York can tell you that there is more to New York than New York City. If you happened to have taken the New York City subway in 2014, you may have noticed advertisements proclaiming this same message. There were the photos of happy people skiing in the Catskills, the boastful statistics about the acreage and breadth of the Adirondack Mountains, and slogans imploring New Yorkers who live in the city to "get out of town," to spend some money vacationing in the further – and, by implication, overlooked – reaches of the vast Empire State. Northern and western New York State can indeed seem remote and removed from, and unnecessary to, the islands of New York City. Beyond the northern highways however, and within the shells of the deindustrialized towns along the old upstate canal routes, there is a far-flung prison archipelago; carceral islands moored within the forests and fields.

Just as there is more to New York State than New York City, there is also more to the city than the city. New York City forms only part of an extensive carceral landscape that encompasses the rural regions to its west and north, part of a re-formed and stretched-out series of relationships – political, infrastructural, financial – that constitute mass incarceration in the Empire State; spatial relationships that enable the sorts of authoritarian policing and repression that have so characterized New York's transition into the twenty-first century. The story of infrastructure and organizing in the northern Adirondacks is but one part of this broader move to mass incarceration in the United States over the past 40 years. Our task as critical scholars and abolitionists, then, is both difficult and clear. We must organize to re-purpose all of this toward liberation, toward a truly social production.

References

Anonymous A (2013) Interview with author, July.
Anonymous B (2014) Interview with author, July.
Besser TL and Hanson MM (2004). Development of last resort: The impact of new state prisons on small town economies in the United States. *Journal of the Community Development Society* 35: 1–16.
Bonds A (2006) Profit from punishment? The politics of prisons, poverty and neoliberal restructuring in the rural American Northwest. *Antipode* 38: 174–177.
Bonds A (2009) Discipline and devolution: Constructions of poverty, race, and criminality in the politics of rural prison development. *Antipode* 41: 416–438.
Boyer G (1986) One man's opinion. *Malone Evening Telegram*, January 3.
Camoin Associates (2009) *Town and Village of Malone Economic Development Plan.* Saratoga Springs, New York.
Caro R (1974) *The Power Broker: Robert Moses and the Fall of New York.* New York: Knopf.
Correctional Association of New York (1982) *The Prison Population Explosion in New York State: A Study of its Causes and Consequences with Recommendations for Change.* New York: Correctional Association of New York, March.
Dao J (1993) Blazing a power trail in the Adirondacks: When Senator Ronald Stafford says not in my district, a whole state goes without. *New York Times*, June 24.
Department of Corrections and Community Supervision (2014) *Facility Listing.* New York State. Available at: www.doccs.ny.gov/faclist.html.
Fraser J (2000) An American seduction: Portrait of a prison town. *Michigan Quarterly Review* 39: 775–795.
Gandy M (2002) *Concrete and Clay: Reworking Nature in New York City.* Cambridge, MA: MIT Press.
Gilmore RW (1998) Globalisation and the U.S. prison growth: From military Keynesianism to post-Keynesian militarism. *Race and Class* 40: 171–188.
Gilmore RW (2002) Fatal couplings of power and difference: Notes on racism and geography. *The Professional Geographer* 54: 15–24.
Gilmore RW (2007) *Golden Gulag: Prisons, Surplus, Crisis, and Opposition in Globalizing California.* Berkeley: University of California Press.

Gish J (2006) The Golden Year. *Times Union*, February 5.

Glynn T (1995) Dannemora: The birth and death of a frontier prison. *Corrections Today* 57: 130–63.

Gramsci A (1971) *Selections from the Prison Notebooks.* New York: International Publishers.

Harvey D (1996) The geography of capitalist accumulation. In Agnew J, Livingstone D, and Rogers A (eds) *Human Geography an Essential Anthology.* Malden: Blackwell.

Harvey D (2006a) Space as a key word. In *Spaces of Global Capitalism.* New York: Verso.

Harvey D (2006b) *Spaces of Global Capitalism: Towards a Theory of Uneven Geographical Development.* New York: Verso.

Harvey D (2007) [1982] *The Limits to Capital.* New York: Verso.

Hooks G, Mosher C, Genter S, Rontolo T, and Laboa L (2010) Revisiting the impact of prison building on job growth: Education, incarceration, and county-level employment, 1976–2004. *Social Science Quarterly* 91: 228–244.

Hooks G, Mosher C, Rotolo T, and Laboa L (2004) The prison industry: Carceral expansion and employment in U.S. counties, 1969–1994. *Social Science Quarterly* 85: 37–57.

Huling T (2002) Building a prison economy in rural America. In Mauer M and Chesney-Lind M (eds) *Invisible Punishment: The Collateral Consequences of Mass Imprisonment.* New York: The New Press.

Jacobs JB and Berkowitz L (1983) Reflections on the defeat of New York state's prison bond. In Jacobs JB (ed) *New Perspectives on Prisons and Imprisonment.* Ithaca, NY: Cornell University Press.

Johnson WO (1974) Back where the Games belong. *Sports Illustrated* 41(19): 28.

Kaplan T (2011) Cuomo administration closing 7 prisons, 2 in New York City. *New York Times*, July 1.

King R, Mauer S, and Huling T (2003) *Big Prisons, Small Towns: Prison Economics in Rural America.* Washington, DC: The Sentencing Project.

Kohler-Hausmann J (2010) "The Attila the Hun Law": New York's Rockefeller drug laws and the making of a punitive state. *Journal of Social History* 44: 71–95.

Krieg F (2013) Hundreds rally against Chateaugay prison closure. *Press-Rupublican*, October 6.

Lawson V, Jarosz L, and Bonds A (2008) Building economies from the bottom up: (Mis) representations of poverty in the rural American Northwest. *Social and Cultural Geography* 9: 737–753.

Lewis WD (1965) *From Newgate to Dannemora: The Rise of the Penitentiary in New York, 1796–1848.* Ithaca, NY: Cornell University Press.

Local Clergy Backs Prison (1978) *Lake Placid News*, September 29, p. 2.

Malone farm sold; may be new shopping center (1989) *Malone Evening Telegram*, January 4.

Marx K (rpt. 1993) *Grundrisse.* New York: Penguin Books.

McDonald D (1980) *The Price of Punishment: Public Spending in Corrections in New York.* Boulder, CO: Westview Press.

McGreevy P (2009) *Stairway to Empire: Lockport, the Erie Canal, and the Shaping of America.* Albany: State University of New York Press.

Millard PJ (1986) Blueprint for tomorrow – Facility planning, design, and construction. *Corrections Today*: 48.

Morrell A (2012a) "Municipal welfare" and the neoliberal prison town: The political economy of prison closures in New York state. *North American Dialogue* 15: 43–50.

Morrell A (2012b) *The Prison Fix: Race, Work, and Economic Development in Elmyra, New York.* New York: The CUNY Graduate Center.

Say no to closures (2013) *Adirondack Daily Enterprise*, August 7.

Schlosser E (1998) The prison-industrial complex. *The Atlantic*, December, pp. 51–77.

Segal J (2012) "All of the mysticism of police expertise": Legalizing stop-and-frisk in New York, 1961–1968. *Harvard Civil Rights-Civil Liberties Law Review* 47: 573–616.

Stafford Papers, M.E. (1975: Series 2, Box 1, Folder 34); (1976: Series 2, Box 1, Folder 38; Series 2, Box 1, Folder 32; Series 2, Box 1, Folder 29; Series 1, Box 1, Folder 10); (1977: Series 2, Box 1, Folder 19); (1985: Series 3, Box 3, Folder 2). Grenander Department of Special Collections and Archives at Albany, State University of New York.

Thompson M (2005) Former state Sen. Ronald Stafford dies. *Post Star.* June 25.

Wacquant L (2009) *Prisons of Poverty.* Minneapolis: University of Minnesota Press.

Williams R (1977) *Marxism and Literature.* New York: Oxford University Press.

11 From prisons to hyperpolicing

Neoliberalism, carcerality, and regulative geographies

Brian Jordan Jefferson

One might assume that substantially reducing the length of prison sentences would effectively dismantle this new system of control. That view, however, is mistaken.

(Michelle Alexander (2010: 93–94))

The spatial practice of a society secretes the society's space.

(Henri Lefebvre (1992: 38))

Introduction

The work of carceral geographers has been foundational in demonstrating historical linkages between neoliberal economic restructuring, uneven development, and eruptions of prison spaces in the US over the past three decades (Bonds, 2013; Che, 2005; Gilmore, 2007; Loyd *et al.*, 2012; Martin and Mitchelson, 2008; Moran *et al.*, 2013; Mountz, 2012; Mountz *et al.*, 2012; Peck, 2003). This rich and diverse scholarship has emerged in response to the colossal ethnoracialized prison buildup during this period, which many have shown as a neoliberal response to managing superfluous labor fractions in a deindustrializing, post-Keynesian economy (Bonds, 2009; Che, 2005; Gilmore, 2007). Through detailed chronicling of jails, prisons,[1] and detention centers, scholars in this emerging field have cast light on how these institutions serve as key spatial forms by which marginalized populations have been contained and discarded by recent political economic mutations, and as such they have played a key role in historicizing mass incarceration in the United States.

But in New York City, the regulation of precarious ethnoracialized labor has taken on its own distinct trajectory in recent decades. Unlike most major US cities such as Milwaukee, Houston, Los Angeles, and Chicago, New York City's recourse to policing as a means of enveloping and marginalizing discarded wage-laborers has overshadowed its reliance on enclosed correctional facilities. In fact, between 1990 and 2000, the number of full-time employees working in New York City's police department increased by 23 percent, nearly double the rate of any other major US city during this period (Reaves and Hickman, 2002). Conversely, from 1991 to 2009, the city's jail population *decreased* by 40

percent (NYSCJPTF, 2013), and New York State was one of only two in the US
to see its jail population shrink (Stephan and Walsh, 2011).

In this chapter I argue that during two administrative junctures near the turn
of the century, the New York Police Department (NYPD) supplemented imprisonment by carceralizing public spaces via *hyperpolicing*, a distinct mode of law
enforcement rendering targeted areas more "prisonlike" inasmuch as it performed three of the neoliberal prison's core socio-spatial functions: marking ethnoracialized deskilled labor with criminal identities, mapping new regulative
geographies of the state, and managing mobilities through a spate of new disciplinary techniques. Reconstruction of the historical emergence of this distinct
species of carcerality is useful in an anticipatory capacity, particularly in light of
several US state-level governances contemplating mass releases of incarcerated
persons due to post-recession austerity measures. Analyzing these historical
blocks thus assists researchers in forecasting carceral forms "after the war on
crime" (Frampton *et al.*, 2008), and developing a critical lens in anticipation of
prospective modes of carceral governance.

Carceral geography and transcarceralization

Geographers have fielded a rich body of literature chronicling the scope with which
the neoliberal carceral system produces new subjectivities and spaces, and have
illustrated the uneven effects it has throughout the broader social body (Allsprach,
2010; Baer and Ravneberg, 2008; Dirsuweit, 1999; Moran, 2014). On the one hand,
scholars demonstrate how prisons and detention centers are robust engines of sub-
jectivity- and spatial-production, detailing three core functions of neoliberal carceral
practice: marking deskilled labor fractions with criminal identities (Allsprach, 2010;
Bonds, 2013; Moran, 2013, 2014); mapping new regulative geographies of the state
(Gilmore, 2007; Loyd *et al.*, 2012; Martin and Mitchelson, 2008; Mountz, 2012;
Mountz *et al.*, 2012; Peck, 2003); and managing the behaviors and mobilities of targeted subjects with ever-intensifying techniques (Baer and Ravneberg, 2008; Dirsuweit, 1999; Milhaud and Moran, 2013; Moran, 2012; Morin, 2013; Sibley and van
Hoven, 2009; van Hoven and Sibley, 2008). This research casts into sharp relief the
ways in which this tripartite process extends beyond sites of correctional and detention facilities, and insinuates itself into "transcarceral" spaces that are not fully
inside, nor fully outside, the broader carceral complex (Allsprach, 2010; Moran,
2013, 2014; Shabazz, 2009). While these literatures document how carceral regulation overflows the walls of prisons and detention centers, the role police play in dif-
fusing carceral space receives less attention.

Works that focus on transcarceral spaces first demonstrate that prisons and
detention centers are characterized by vigorous discursive production, anchored
in *marking* human beings with stigmas that resonate far beyond carceral institutions (Allsprach, 2010; Loyd *et al.*, 2012; Moran, 2014; Nevins, 2012). Ground-
ing inquiry in political economy and uneven spatial development, a variety of
studies illustrate the proliferation of prisons and detention centers as being
anchored in a broader strategy of branding criminal identities onto superfluous

wage laborers alongside deindustrialization, the global foot-loosing of capital, and disintegration of social safety nets (Bonds, 2009; Peck, 2003). Within this political economic matrix, carceral institutions inscribe a myriad of devalued subjectivities including those pertaining to ethnicity/race (Bonds, 2006, 2009; Shabazz, 2009), gender/sexuality (Allsprach, 2010; Dirsuweit, 1999; Moran, 2014), nationality (Martin and Mitchelson, 2008; Mountz *et al.*, 2012), regionality (Bonds, 2009, 2013; Che, 2005), and age (Wahidin, 2006).

The disciplinary effects these inscriptions have on those processed through prisons shadow individuals after release, drip-feeding carcerality throughout civil society. Allspach (2010) thus develops the concept of transcarceral space to engage the "widened web" of neoliberal inflected prison control of formerly incarcerated women in Canada. Intensively surveilled halfway houses, invasive parole stipulations, and structural restraints via responsibilized social services all combine to impose considerable regulations on the habituses, mobilities, and in some cases mortality of individuals after release. Likewise, Moran (2014) demonstrates how prison stigmatization is grafted onto the body, subsequently regulating the behaviors, mobilities, and subjectivities of formerly incarcerated persons. Here crude dental care; tattoo culture; and linguistic, drinking, and smoking habits characteristic of Russian prisons leave permanent and semi-permanent inscriptions upon the formerly incarcerated body, significantly lengthening the carceral nexus across space and time.

Second, carceral geographers also chronicle how the production of prisons and detention centers is implicated in *mapping* new regulative geographies of neoliberal governance. "Regulative geographies" refers to those spaces, behaviors, persons, and issues over which governances posit jurisdiction and endeavor to manage through discourse and/or material practice. On a vertical axis, those focusing on top-down/bottom-up relations of governance illustrate how the proliferation of prisons interface broader morphologies in the neoliberal state like the downstreaming of select regulatory functions to lower administrative tiers and the private sector. Prison building, it is often noted, is imbricated in wider programs devolving the responsibility of local economies, job creation, and social protections to municipal and private sectors (Bonds, 2006, 2009; Che, 2005; Gilmore, 2007). On a horizontal axis, those concentrating on transformations occurring across civil society examine how prisons and detention centers are deeply engaged in remapping borders that demarcate socio-spatial difference, producing new hybrid spaces neither inside nor outside civil society. These works have been especially important in examining detention centers' new roles in delineating, and sometimes obscuring, national borders and the parameters of civil society (Mountz, 2012; Mountz *et al.*, 2012). These borders are shown to be highly unstable and pregnable, generative of ambivalent heterotopic spaces resistant to sharply defined inside/outside binaries (Baer and Ravneberg, 2008; Dirsuweit, 1999; Moran, 2013). Characterized by flows of sentenced persons, families of the incarcerated, material goods, communication, and experiences, these indistinct, liminal spaces reflect the new spatial configurations and disfigurations engendered horizontally across civil society via carceral governance.

Third, geographical researchers investigate the material practices employed inside prisons and detention centers aimed at *managing* the behaviors and mobilities of the interned, which are in turn replicated in public space. Such works draw upon Martin and Mitchelson's (2008) description of carceral management as the mobilization of violence to "hold human beings without consent," and cast light on the shadowy practices embedded in prisons including surveillance tactics, sensory deprivation, torture, and a range of other techniques in attempts to construct prison time-space (Gregory, 2006; Jared and Lee, 2006; Sibley and van Hoven, 2009; Valentine and Longstaff, 1998; van Hoven and Sibley, 2008). Milhaud and Moran (2013) show how these practices work in tandem with architecture, illustrating how both compartmentalized and communalized carceral regimes combine management practices and physical space in unique ways to label, segregate, and/or cluster prisoners according to different legal categories and disciplinary objectives. Shabazz (2009) provides a unique study of how such practices are exported into public areas so as to constitute the "prisonization of black quotidian spaces," interrogating how carceral forms organize the living and working spaces of the racialized poor in Chicago and South Africa. Housing projects and mining compounds are constituted via periphrastic spatial orders wherein bodies are crammed together, food distribution is substandard, and space is coercively arranged and managed – especially by police. While Shabazz mentions the police's role in producing the carceral milieux in urban landscapes, details about this role in discursively marking racialized labor, remapping the regulative terrain of the state, and micromanaging human flows and behaviors require further exploration.

Investigating policing and the carceralization of cityspace

Two main questions underscore analysis of the police function in the carceralization of cityspace: (1) To what degree does policing facilitate marking ethnoracialized deskilled labor fractions with stigmatized identities, mapping new regulative geographies of the state, and managing the habituses and mobilities of raced, classed, and gendered subjects? (2) To what degree are these practices impressed upon those who have never set foot in prisons? To be sure, several urban scholars elucidate functional resemblances between prisons and ghettoized enclaves, illustrating how the latter serve as "warehouses" for the social reproduction of criminalized, ethnoracialized, precarious labor (Parenti, 2008; Peck and Theodore, 2008; Wacquant, 2000, 2001; Wilson, 2005, 2006). Moreover, many demonstrate how urban police departments have employed a spectrum of adrenalized practices over the past two decades to execute a horizontal "symbolic mission" (Wacquant, 2009) of partitioning ghettoized spaces, and enforcing the border between mainstream society and the urban periphery (Davis, 1990; Smith, 2001). Within this spatial schema ghettoized sections function as repositories of persons marked by the prison system, their communities thereby trapped within the orbit of systemic incarceration (Peck and Theodore, 2008). However, incarcerated and formerly incarcerated subjects' first point of contact with the carceral system is with the police, who, in addition to the court system, play a constitutive role in marking them as criminal as well.

With rare exceptions (Herbert, 1996; Jefferson, 2014; Manning, 2003; O'Malley and Palmer, 1996), scholars interpret policing solely as a matter of physically managing persons, with scant attention paid to its discursive capacities in actively marking humans and producing spaces. But examination of the NYPD's morphology during recent history reveals a much more comprehensive socio-spatial function than simply managing the mobilities of stigmatized dislocated sub-populations. In the course of the past two city regimes, the NYPD has adopted all three socio-spatial functions of neoliberal carceral institutions, coalescing in what may be termed a *hyperpolicing*: hyper in the degree to which it exceeds the law enforcement function and categorizes behaviors and persons as "disorderly"; in its hyperrealist approach constituting new virtual spaces; and in the sense of its unprecedented activity in the minutiae of quotidian life. These three socio-spatial functions interface seamlessly with broader carceral processes, and have been central in reorganizing cityspace during the past two decades.

Thus I chronicle the development of New York hyperpolicing in two discernible yet overlapping administrative junctures: neoliberal restructuring under Mayor Giuliani's administration (1994–2001), and high-octane entrepreneurialism under Mayor Bloomberg's administration (2002–2013). I examine the first stage, characterized by frenetic discursive activity and vertical state rescaling, through data from the Archive of Rudolph W. Giuliani in New York's Municipal Reference Library (henceforth "Giuliani Archive"). Here I examined 30 folders from the "subject files," "department correspondence," and "projects/subjects" subseries, each consisting of 50 to 250 files. I investigate the second stage, marked by horizontal differentiations of space and intensified managing techniques, through data from 452 documents from the city government's digital Office of the Mayor archive (henceforth "Bloomberg Archive"). In each document I singled out and analyzed how both regimes situated the police department in producing and inscribing stigmatized subjectivities, reconstituting the regulative terrain of state intervention, and managing behaviors and mobilities of the ethnoracialized poor.

Neoliberal restructuring and the formation of hyperpolicing

While there is a profusion of insightful scholarship that locates Giuliani-era (1994–2001) NYPD reforms within the political economic matrix of neo-liberalism, the deep connections these transformations shared with broader carceral developments is only casually considered. But the NYPD's metamorphosis alongside the administration's retrenchment of the public sector, deregulation of financial and commercial markets, and obsessive rhetoric about the individual responsibilities of the inner-city poor, involved the implementation of several socio-spatial functions emerging throughout the US prison system. In particular, the police and public discussions about policing were utilized as focal instruments for marking stigmatized identities onto ethnoracialized labor fractions dislodged by the city's economic tailspin. In addition, the NYPD's regulative landscape morphed in ways resembling wider carceral governance, as it devolved decision-making powers to lower administrative tiers and extended its disciplinary reach into new dimensions of civil society. These dynamics situated policing at the heart of the state's role in uneven social reproduction and accommodating real estate capital in the emerging economic climate, advancing significantly the socio-spatial logic of carcerality in targeted areas in the process.

Hyperpolicing and discursive production

During New York City's 1993 mayoral campaign, debates about policing constituted a discursive field on which the urban indigent were marked as sources of environmental degeneracy and criminal profligacy. On this symbolic terrain, discussions about NYPD reform were crucial to inscribing deviant and criminal subjectivities onto those economically marginalized by the city's loss of industrial jobs, the financial crisis, and planned shrinkage packages. Deploying debates about policing and the police department to stigmatize the economically dislocated was catalyzed by an internal contradiction of capital festering since the 1973 stock market crash, which generated a bloated population of ethnoracialized underemployed, unemployed, and visibly homeless persons across the cityscape. Since the late 1970s, the city's blueprint for neutralizing recession dislocations revolved around making the city more accommodating to outside investors, developers, and tourists. Yet this blueprint pivoted on large-scale targeted disinvestments in public infrastructure, prompting such widespread visible social and physical disorder that the city was an unattractive location to outside capital. These dynamics gained considerable momentum after the 1987 stock market plunge, which prompted various cost-cutting measures resulting in a 21,000-person reduction in the public workforce in 1990 to 1991. In addition, more industry jobs were lost in New York during this stretch than in any other major American city, save Philadelphia. Middle-wage jobs began a ten-year decline (3.7 percent), while low-wage jobs increased by 6.4 percent (Adler, 2002).

The social dislocations wrought by these structural disruptions materialized in profusions of visible indigence and disorder (see Vitale, 2008). By 1989 an estimated 10,000 mentally ill people found themselves homeless, and 11,000 adults and 5,000 families lived in homeless shelters (Finder, 1987). By the early 1990s these numbers had multiplied, with estimates of up to 50,000 homeless people at one point, and 200,000 ill-housed persons roving from shelter to shelter (Wackstein, 1992). By late 1992, despite improvements in the national economy, the city's unemployment rate had reached 11.5 percent, its highest since the 1976 financial crisis (US Bureau of Labor Statistics, 2014). That same year, 25 percent of all New Yorkers received incomes falling below the federal poverty line, and 44 percent of children lived at or below the poverty line (Lueck, 1992).

The ethnoracial composition of those bearing the brunt of restructuring was stark. In 1990, of the 1,825,000 New Yorkers who lived below the poverty line, 32.6 percent were Latino, 23.7 percent were Black, and 9.2 percent were White

(Calcagno, 2013). This was in stark contrast to the national scale, in which the impoverished class was 71 percent White, 17 percent Black, and nearly 12 percent Latino the same year (United States Bureau of the Census, 1991). By 1992, 6.6 percent of Whites, 14 percent of Blacks, and nearly 16 percent of Latinos were unemployed (US Bureau of Labor Statistics, 2014). Between 1988 and 1992, homeless adults in the New York City's shelter system were 62 percent Black, 23.5 percent Latino, and 8.3 percent White (Culhane *et al.*, 1998).

From a discursive standpoint, the city's approach to containing these disruptions crystallized in a program of mobilizing public debates about crime and policing to assail dislocated persons through a fusillade of criminological tropes. This strategy took its cue from the broken windows doctrine, the "guiding theory" of police reform (Kelling and Bratton, 1998), which postulated that "disorderly, disreputable, obstreperous and unpredictable" individuals including pandhandlers, drunks, drug addicts, loiterers, and the mentally ill were catalysts of urban decline (Kelling and Wilson, 1982). This sentiment echoed among policy makers, notably Daniel Moynihan, who criticized the authorities for defining fewer types of behaviors as disorderly, and enjoined city officials to reverse the trend amid rising crime rates. This discursive strategy was quickly adopted by candidate Giuliani during the 1993 mayoral campaign, in which he portrayed low-level quality of life offenders as "urban terrorists" and squeegee operators (those washing car windows) as "extortionists." The homeless were routinely regarded as dependants of the state, draining public expenditures and deteriorating the city's quality of life with their mere presence. Illegal street vendors were said to be threats to small business retarding recession recovery. Giuliani marked these persons and groups as "symbols of disorder" en masse, announcing that low-level disorders would be met with zero tolerance enforcement.

Moreover, the NYPD's role in marking the socioeconomically vulnerable as agents of deviance and criminality involved what Neil Smith (1998) describes as a "visceral identification of the culprits," repositioning the department as a discursive apparatus in labeling stigmatized subjects. In 1994 the NYPD released its Police Strategy No. 5 to "reclaim public spaces" throughout the five boroughs (Giuliani and Bratton, 1994). This was intended to increase the public use of space by aggressively targeting "lower grade" criminal activity such as peddling, panhandling, squeegee cleaning, street prostituting, playing loud boomboxes, driving noisy motorcycles, participating in loud parties, graffiti writing, and being dangerously mentally ill in the street. The department thus marked several "low-level criminals" as threats to the public environment: illegal vendors, beggars, loiterers, visibly mentally ill people, visibly homeless people, streetside sex workers, truants, reckless bikers, turnstile jumpers, peep-show performers, and a range of petty hucksters. It coded these persons as "symbols of disorder" and "visible signs of a city out of control, a city that cannot protect its space" (Giuliani and Bratton, 1994). In mid-March 1994, the NYPD began to observe the disorderly conduct statute with renewed vigor, and, in accordance with §240.20 of New York State's penal law, issued disorderly conduct tickets for a variety of behaviors, including unreasonable noise, obscene language, and disorderly gestures.

Remapping regulative geographies

In addition to utilizing policing debates and the police department as discursive sluices for marking the ethnoracialized poor as agents of disorder, early phases of the city's ever-neoliberalizing governance involved deep mutations in the state's regulative geographies. On the vertical axis, the hyperpolicing regime pivoted on devolving authority down the institutional hierarchy, and establishing the precinct as the focal territorial unit of policy formulation. Downstreaming authority occurred most notably through its 1994 "re-engineering" program, which prompted streamlining the department's bureaucratic field and empowering lower institutional strata (Bratton, 1998; Silverman, 1999). These transformations were characterized by a process of decentralization that involved relocating decision-making authority over daily ground operations, planning, staffing, and implementing anti-crime initiatives from NYPD headquarters to Patrol Borough and Precinct Commanders. Commissioner Bratton based devolving authority to this middle stratum in accordance with their familiarity with community councils and local organizations. He explained in a 1996 speech, "Decentralizing and Establishing Accountability," that a decentralized NYPD meant that "[precinct commanders] were empowered to assign officers as they saw fit, to focus on the priorities of the neighborhood. Whatever was generating the fear in their precinct, they were empowered to address it by prioritizing their response" (Henry, 2003).

Vertical institutional rescaling meshed with several horizontal transformations, namely those affording police precincts new powers in conceptualizing and organizing cityspace. Indeed, the department's devolutionary initiatives produced new virtual spaces, as they partitioned the cityscape into eight "patrol boroughs" with two to three divisions and precincts inhabited by approximately 100,000 persons apiece (Bratton, 1998; Nagy and Podlony, 2008). Ceaseless flows of crime data were also compiled into commander profile reports and report cards of non-crime management, including patrol and investigative methods, establishing a quota system that measured commanders' performance by hard statistics and complex data analysis. In this sense, the new system implemented a results-oriented, 'bottom-line' approach that reimagined communities as targets, and escalated the predatory ethos in high-crime minority precincts (Eterno and Silverman, 2006; Sciarabba, 2009).

The computerization of Compstat's crime pattern mapping system induced a hyperrealist approach to conceptualizing problematic precincts and areas, as the department's capacity to geocode its low-threshold definition of social disorder increased considerably. Before the NYPD's crime mapping was computerized, precincts teamed up with civilians to generate their own pin maps of mostly drug and violent crimes, including robberies, shootings, grand larceny, and murders. Pin maps were not mandatory and depicted various kinds of crime data on various time-scales from precinct to precinct. As a result of the lack of coordination and uniformity among the precincts, pin maps were used primarily for the perfunctory purpose of providing accessible data to department headquarters. Neither department nor precinct administrators utilized crime mapping on a consistent basis in day-to-day operations for strategic purposes (Henry, 2003; Silverman, 1999).

However, as Compstat computerized the department's cartographical praxis, two developments occurred, advancing its role in carceralizing targeted precincts. On the one hand, the more computerized Compstat became, the lower the level of crimes it began tracking. Once the department integrated the radar-like MapInfo94 software, Compstat added additional layers of public behaviors to its digital maps. Not only did it map the seven major crimes on the UCR Index, it also added geocoded layers depicting civilian complaints, daily summons tallies, parolee residences, and desk appearance tickets (DATs) for minor violations and misdemeanors, including public consumption, littering, and pandhandling. On the other hand, with this new data aggregate at its disposal, the department came to view Compstat as a predictive technology to "determine and geographically depict the time and location crimes are *most likely* to take place" (Henry, 2003, emphasis added). No longer did clustered criminal activity prompt increased police deployments; now certain statistical profiles, not necessarily inclusive of actual crimes committed, designated an area subject to intensive policing. Armed with this more statistics-heavy map of New York City precincts, in 1994 the NYPD parceled up the cityscape into 24 precincts with quality of life problems, 29 precincts with prostitution epidemics, and 251 locations wracked by "metastasized disorder," marked for "saturation patrols" involving targeted deployments of 1,000 officers (Giuliani Archive, 1994).

This new institutional landscape constituted a novel field of regulative intervention, characterized by what the commissioner deemed "miniature police departments" responsible for employing "grassroots creativity" when developing anticrime strategies (Nagy and Podlony, 2008). One on-ground permutation involved the NYPD's role in carceralizing disadvantaged schools. In 1998 the NYPD took over enforcement duties outlined in school disciplinary codes via a joint safety agreement between the Board of Education and police department, and established the School Safety Division (SSD). The SSD consisted of nine patrol borough commands headed by borough managers in charge of safety and security measures inside and around school vicinities. Furthermore, the division increased school safety agents and armed school police officers stationed at public schools from 3,200 to 5,200, encouraging them to treat school buildings and grounds as their beats (Vera Institute, 2001). As the joint security agreement authorized the NYPD to enforce schools' disciplinary codes, safety agents were not only responsible for policing violent and criminal activity, but also for regulating minor misbehaviors such as cell phone use, smoking, and class-ditching via summonses, citations, juvenile detention referrals, criminal charges, and arrests. The new module also established the Truancy Reduction Alliance to Contact Kids (TRACK) to locate and intercept youth caught evading school.

High-octane entrepreneurialism and hyperpolicing development

Mayor Michael Bloomberg (2002–2013) began his tenure promising to maintain his predecessor's crime fighting policy. His assurances overlapped with a decisively "high-octane" (While *et al.*, 2004) economic policy characterized by grand-scale projects to develop center city functions and an ever-centralizing entrepreneurial strategy helmed by the city administration. Enmeshing NYPD policy deeper into the dictates of real estate capital, the administration shifted the way the hyperpolicing regime marked ethnoracialized poverty, particularly in its new-found focus on categorizing entire geographic vicinities as deviant through its "problem people/problem places" discourse. This new administrative juncture advanced the carceralization of targeted spaces through a new zonal strategy that deepened differentiations between mainstream and ethnoracialized localities, and managed the latter with greater invasiveness and intensity.

Zonal regulative geographies

Using policing and debates about policing to define, demarcate, and regulate devalued labor fractions became more geographically articulated during Bloomberg's administration. From his 2002 inaugural speech onward, the mayor announced a "problem people/problem places" strategy as the center-piece of his crime control policy (Bloomberg Archive, 2005). Problem people/problem places was characterized by increasing police deployments in targeted areas by the thousands, which were made responsible for controlling quality of life offenses, including littering, street vending, marijuana smoking, squeegee operating, loitering, playing loud music, reckless driving, and writing graffiti. As such, violent criminals, chronic misdemeanants, drug offenders, shoplifters, prostitutes, truants, rowdy students, and disorderly youth were marked under the same category of "problems," and the administration turned to Compstat to locate and manage clusters of these persons through immersive patrolling (Bloomberg Archive, 2004).

Densities of quality of life offenders such as the 75th in East New York, 83rd in Bushwick, and 60th in Coney Island were marked as "problem places," in which the administration installed "Impact Zones" consisting of hundreds of extra officers performing special operations to "prevent crime before it happens" (Bloomberg Archive, 2005). Impact zoning was introduced in the form of the 2002 Operation Clean Sweep, which involved the asymmetrical deployment of 1,500 police in 24 "pockets of the city" to increase issuance of summonses for squeegee operating, panhandling, unlicensed peddling, public alcoholic consumption, marijuana smoking, public urination, and homeless encamping. In its first year, Clean Sweep generated over 20,000 arrests and 209,000 summonses. In 2005, it was credited for 33,000 arrests and 350,000 summonses. In this vein, the department unleashed Operation Spotlight which concentrated on tracking petty misdemeanants; Operation Silent Night which involved an anti-noise enforcement; Operation Safe Housing which intensified NYPD patrolling of public housing complexes; and expanded the number of NYPD policed educational institutions via "Impact Schools" involving the "aggressive enforcement by police officers and school safety agents against all forms of student disorder." Reflecting on its first two years, Bloomberg lauded Impact Zones for generating 72,000 arrests (Bloomberg Archive, 2004).

Impact zoning engendered a recursive feedback loop between zonal management and zonal stigmatization. That is, the more intensely these geographic "pockets" were policed, the more instances of disorder were identified, and thus the more disorderly the data suggested those areas were (Jones-Brown *et al.*, 2010; Jones-Brown and King-Toler, 2010). Thus, even though crime declined across these precincts in 2003, impact zone policing was extended to 11 additional precincts and disadvantaged "Impact Schools" the following year. This extension was correlative with the NYPD's growing reliance on automated information-gathering, data-analysis, and surveillance systems. Bloomberg asserted the regulative geographies established through problem people/problem places combined with crime data to create an:

> effective law enforcement [that] is all about pattern recognition. And the hallmark of the next phase of our crime-fighting efforts will be the increasingly sophisticated use of information technology and DNA evidence. To identify and stop emerging crime patterns before they become crime waves, and to ensure that dangerous criminals who might otherwise go uncaught and unpunished face justice.
>
> (Bloomberg Archive, 2005)

Another spatially targeted program launched by the Bloomberg administration was the "vertical patrol" program, which, drawing from the 1991 Operation Clean Halls, deployed police in at least 3,895 low-rent private housing complexes. Clean Halls adhered to the New York City Housing Authority's 'do-not-enter' bylaws, which prohibit people not residing in or visiting a resident of city housing complexes from entering. According to the department's vertical patrol policy, officers were instructed to stop trespass suspects and ask if they lived in the building, were visiting a resident, or had specific business in the building. Officers were further directed to request identification or keys to the building from the detainees in question. Between 2006 and 2010, the NYPD conducted 329,446 stops on suspicion of trespassing, only 7.5 percent of which resulted in an arrest, and 5 percent in summons. In 2010 the ten precincts with the most stops totaled almost the entire number of stops in the other 66 precincts (NYCLU, 2011). During this period in the Brownsville section of Brooklyn alone, an area with the densest concentrations of public housing in the city, upward of 52,000 stops were executed, equaling about one stop a year per resident. Males between the ages of 15 and 34 in this area were stopped on average five times a year (Rivera *et al.*, 2010).

Impact zoning went hand-in-hand with the viral proliferation of surveillance and data gathering, which prompted new developments in the horizontal ordering of public space. In 2006 the NYPD unveiled its Real Time Crime Center, a 24-hour data bank staffed by officers who provide detectives with instant information on crimes scenes, crime patterns, and criminal suspects (Jefferson, 2011). The Center also houses transcripts of 911 calls and personal information of people with criminal records. Around the same time the administration launched its "Ring of Steel" and Manhattan Security Initiative surveillance programs, consisting of thousands of closed-circuit "crime cameras" in select neighborhoods, stationed strategically according to crime pattern mapping. City officials teamed with the private sector to implement the Security Initiative to further develop this infrastructure with license plate readers and helicopters equipped with cameras capable of facial and movement recognition from two and 12 miles away, respectively. The department also partnered with Microsoft to integrate a vast web of cameras, license plate readers, and criminal databases into the Domain Awareness System (DAS), a centralized control center enabling NYPD personnel to access footage from cameras within 500 feet of 911 reports of suspicious activity (Dolmetsch and Goldman, 2012). Within the first four years of problem people/problem places, the number of surveillance cameras increased four-fold (Siegel *et al.*, 2006).

Surveillance technologies performed a key function in differentiating problem areas from those of mainstream economic and social activities. Indeed, economically distressed ethnoracialized localities were strewn with highly visible crime security cameras, most often mounted on posts with blue flashing lights. These conspicuous devices were emplaced with intent to discourage and monitor trespassing, vandalism, and more serious criminal activity. In neighborhoods like Bedford-Stuyvesant, Crown Heights, Jackson Heights, and Jamaica, surveillance cameras were meant to function as crime deterrents as much as they were investigative tools. Crime cameras were gradually adopted by the Housing Authority and have been stationed in 85 high-crime housing developments. The police and housing department have also deployed a small and mobile fleet of 25-foot-tall SkyWatch towers and Mobile Utility Surveillance Towers in neighborhoods experiencing concentrated waves of crime and public events, to keep ostentatious watch over civilians.

But surveillance in the business and financial districts was more inconspicuous, as monitoring devices were attached to existing structures in the built environment, and used more strictly for information-gathering and investigative purposes. Thus, in addition to cameras planted in helicopters, the city installed thousands of CCTVs on skyscrapers' roof-tops, lamp-posts, and traffic signs (Jefferson, 2011). Moreover, in the center city, license plate readers were fixed on the trunks of squad cars and on metal beams that parallel electrical cables. By the turn of the decade, cameras were extended to traffic lanes on all bridge and tunnel entry points to Manhattan, creating an invisible moat that traces flows to and from centers of concentrated capital, without the aggressive conspicuousness found in low-income Black and Brown enclaves.

Hypermanagement of targeted zones

The management of marked bodies within impact zones accelerated under the problem people/problem places regime, particularly in relation to the use of the stop, question, and frisk (SQF) tactic. While SQF data are unavailable for the Giuliani administration, there is a consensus that its use expanded exponentially during Bloomberg's tenure, and came to assume a primary role in managing the behaviors and mobilities of those situated inside problem places (Fagan, 2010; Levine, 2011; NYCLU, 2014; Vitale, 2008). SQF involves patrol officers halting "furtive" persons throughout the public sphere, and, during Bloomberg's tenure, increased by an average of 60,853 additional stops each year. Of the 4,889,838 SQFs performed between 2002 and 2012, 54 percent have been conducted on Blacks and 32 percent on Latina/os (NYCLU, 2014). At around the same time, officers used force during 27 percent of stops involving Latinos, 25 percent of stops involving Blacks, and 19 percent of those involving Whites (Fagan, 2010). By 2010, of the 20 precincts with the highest number of SQFs, three had majority White populations, six had majority Latino/a populations, and 11 had majority Black populations (Bloch *et al.*, 2010).

The carceral dimension of SQFs came into sharp focus through the Impact Schools program, which raised the total headcount of NYPD personnel stationed in educational institutions to approximate the entire police forces of Washington, DC, Detroit, and Seattle, constituting the tenth largest law enforcement agency in the nation. Impact Schooling expanded the SSD disciplinary functions, particularly throughout schools located in impoverished ethnoracialized areas. All but two Impact Schools were located in poor, majority Black, or Latina/o zip codes (see Figure 11.1). The median household income of zip codes with Impact Schools was US$42,670, compared to US$50,657 citywide in 2011, save the two majority-White, Manhattan-located Norman Thomas and Washington Irving high schools (see Figure 11.2). Furthermore, excluding those same schools, the ethnoracial composition of Impact School zip codes was 13 percent more Black, 4 percent more Latino, and 8 percent less White than the city average. The discursive capacities of marking subpopulations with stigmatized identities in these schools are considerable, specifically in their role in reproducing ethnoracialized and class-based youth delinquents. By 2012 Black students accounted for 63 percent of all school-related arrests and summonses, although they make up 23 percent of the overall student population. In comparison to Whites, Black students are 14 times more likely to be arrested, and Latina/o students are five times more likely (NYCSCPTF, 2013). Some 99,000 students enter school through airport-style metal detectors: 82 percent are Black or Latina/o (NYCLU, 2007).

This disciplinary management of ethnoracialized schools works in tandem with the prison system, illustrating the extent to which the carceral web is woven into the fabric of public space. By 2005, the majority of incarcerated youth in New York City were convicted of such non-violent, low-level offenses (CANY, 2006). The 31,879 suspensions in the 2002 to 2003 school year increased to 73,943 in 2008 to 2009, despite a 70,000 student decrease in the overall student

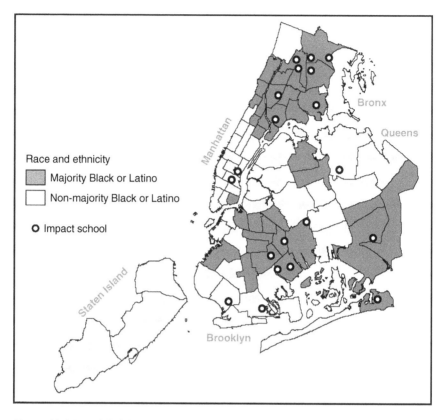

Figure 11.1 Race/ethnicity and Impact Schools, 2010. Map courtesy of Kenneth French.

population during this period (NYCLU, 2011). By the 2011 to 2012 school year, there were four school-related arrests and seven school-related summonses per day. The majority of these were for minor misbehaviors: 70 percent of all summons issued were for "disruptive behavior," and 64 percent were for disorderly conduct, including loud noise, profanity, or loitering (NYCSCPTF, 2013).

Historical geographies of hyperpolicing and the new reconstruction

The historical reconstruction of NYPD hyperpolicing during the eras of neoliberal restructuring and high-octane entrepreneurialism is useful not so much in the sense of "weaving inspirational stories" (Blake, 1999) as in forecasting and critiquing future carceral forms should mass imprisonment crumble under the strain of its own inefficiency. Indeed, advocates of what I term hyperpolicing have already argued that the NYPD's hyper-interventionist approach is a desirable alternative to

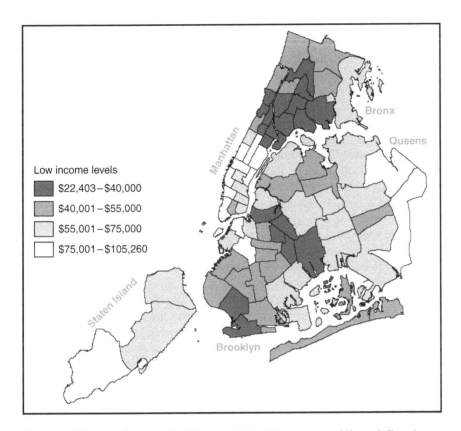

Figure 11.2 Low median household income, 2012. Map courtesy of Kenneth French.

imprisonment, greatly benefiting Black and Latino males (Smith, 2012). Thus Wilson (2006: 99) acutely postulates that if the prison apparatus "evaporates entirely" cities might then confront a scenario wherein the:

> policing of black ghettos and public spaces will probably intensify. Denied unfettered use of prisons to stash, isolate, and mark this population, such a compensatory response would be anything but surprising. That would mean more drug raids, more curfew and no standing ordinances, more intense surveillance of streets and public spaces, and more targeted auto stoppings in cities.

What the hyperpolicing case reveals is that carcerality is a considerably fungible process characterized by a distinct socio-spatial code cast deeply in the logic of neoliberal governance. This code is in no way reducible to prisons and detention centers, and the retrenchment and/or disintegration of these institutions does not equate to the disappearance of carceral space. This is particularly relevant in light of the looming "New Reconstruction" (Frampton *et al.*, 2008), or re-entry crisis born of early prisoner release programs prompted by recent austerity measures across state-level governances (see also Clear, 2007). Indeed, upward of 10,000 inmates are released from state and federal prisons on a weekly basis, 650,000 annually. In California the *Brown v. Plata* (2011) ruling resulted in the court's release order of between 38,000 and 46,000 inmates. In Oregon, where the Department of Corrections accounted for 53 percent of the public safety spending, 2011 budget cuts have prompted plans to close down six minimum-security prisons, reducing its prisoner population by almost 20 percent. While the disintegration of the towering prison state would be a welcome development, the fact that it will prompt influxes of ethnoracialized, stigmatized, low-skilled workers into dysfunctional labor markets poses a host of questions for carceral geographers moving forward.

In this sense, the hyperpolicing case signals the need for further developing decentered ways of identifying and articulating the production of carceral space. It is important to note that further study of these polymorphous spaces requires ethnographic inquiry into the subversive practices that also play a part in shaping the contours of these areas, and how vulnerable ethnoracialized and gendered wage earners experience these areas. Such investigation will serve to broaden and nuance understandings of carceral space, and the discursive and material processes that envelop and marginalize those dislocated by political economic restructuring. It will also provide usable narratives to mediate modulations in carceral governance, and develop alternative modes of crime control, criminal justice, and approaches to producing urban space.

Acknowledgment

I would like to thank Kenneth French, University of Wisconsin-Parkside, Department of Geography, who provided the maps for this chapter.

Note

1 In the US, prisons are designed for long-term incarceration and managed by state- or federal-level government. Jails, on the other hand, perform short-term incarceration and are administered at the municipal scale.

References

Adler M (2002) Why did New York workers lose ground in the 1990s? *The Regional Labor Review* 5: 31–35.

Alexander M (2012) *The New Jim Crow: Mass Incarceration in the Age of Colorblindness*. New York: The New Press.

Allsprach A (2010) Landscapes of (neo-)liberal control: The transcarceral spaces of federally sentenced women in Canada. *Gender, Place and Culture: A Journal of Feminist Geography* 17: 705–723.

Baer LD and Ravneberg B (2008) The outside and inside in Norwegian and English prisons. *Geografiska Annaler: Series B, Human Geography* 90: 205–216.

Blake CN (1999) The usable past, the comfortable past, and the civic past: Memory in contemporary America. *Cultural Anthropology* 14: 423–435.

Bloch M, Fessenden F, and Roberts J (2010) Stop, question and frisk in New York neighborhoods. *New York Times*, July 11.

Bloomberg M (2014) Archives of the 108th Mayor of New York City. Available at: www1.nyc.gov/office-of-the-mayor/archive/index.page.

Bonds A (2006) Profit from punishment? The politics of prisons, poverty and neoliberal restructuring in the rural American Northwest. *Antipode* 38: 174–177.

Bonds A (2009) Discipline and devolution: Constructions of poverty, race, and criminality in the politics of rural prison development. *Antipode* 41: 416–438.

Bonds A (2013) Economic development and relational racialization: "Yes in My Backyard" prison politics and the racialized reinvention of Madras, Oregon. *Annals of the Association of American Geographers* 103: 1389–1405.

Bratton W (1998) *The Turnaround: How America's Top Cop Reversed the Crime Epidemic*. New York: Random House.

Calcagno J (2013) *Trends in Poverty Rates Among Latinos in New York City and the United States, 1990–2011*. New York: Center for Latin American, Caribbean and Latino Studies.

Che D (2005) Constructing a prison in the forest: Conflicts over nature, paradise, and identity. *Annals of the Association of American Geographers* 95: 809–831.

Clear T (2007) *Imprisoning Communities: How Mass Incarceration Makes Disadvantaged Neighborhoods Worse*. New York: Oxford University Press.

Correctional Association of New York (CANY) (2006) *Youth Confined in OCFS Facilities*. New York.

Culhane D, Dejaowski E, Ibanez J, Needham E, and Macchie I (1998) Public shelter admission rates in Philadelphia and New York City: The implication of turnover for sheltered population counts. *Housing Policy Debate* 5: 107–140.

Davis M (1990) *City of Quartz: Excavating the Future of Los Angeles*. London: Verso.

Dirsuweit T (1999) Carceral spaces in South Africa: A case study of institutional power, sexuality and transgression in a women's prison. *Geoforum* 30: 71–83.

Dolmetsch C and Goldman H (2012) New York, Microsoft unveil joint crime-tracking system. *Bloomberg*, August 8. Available at: www.bloomberg.com/news/2012-08-08/new-york-microsoft-unveil-joint-crime-tracking-system.html.

Eterno J and Silverman EB (2006) The New York City police department's Compstat: Dream or nightmare? *International Journal of Police Science and Management* 8: 218–231.

Fagan J (2010) *Floyd, et al.* v. *City of New York, et al: Supplemental Expert Report*. New York: Center for Constitutional Rights.

Finder A (1987) Board approves plan for shelters. *New York Times*, July 20.

Frampton ML, López IH, and Simon J (eds) (2008) *After the War on Crime: Race, Democracy, and a New Reconstruction*. New York: New York University Press.

Gilmore R (2007) *Golden Gulag: Prisons, Surplus, Crisis, and Oppression in Globalizing California*. Berkeley: University of California Press.

Giuliani R Archives of Rudolph W Giuliani (1994–2001) New York: New York Municipal Reference Library.

Giuliani R and Bratton W (1994) *Police Strategy No. 5: Reclaiming the Public Spaces of New York*. New York: Office of the Mayor.

Graham M and Shelton T (2013) Geography and the future of big data, big data and the future of geography. *Dialogues in Human Geography* 3: 255–261.

Gregory D (2006) Vanishing points: Law, violence and exception in the global war prison. In Gregory D and Pred A (eds) *Violent Geographies: Fear, Terror, and Political Violence*. New York: Routledge, 205–237.

Henry V (2003) *The Compstat Paradigm: Management Accountability in Policing, Business and the Public Sector*. Flushing, NY: Looseleaf Law Publications.

Herbert S (1996) *Policing Space: Territoriality and the Los Angeles Police Department*. Minneapolis: University of Minnesota Press.

Jared S and Lee E (2006) Figuring the prison: Prerequisites at Abu Ghraib. *Antipode* 38: 1005–1022.

Jefferson BJ (2011) Full spectrum surveillance: NYPD, panopticism and the public disciplinary complex. In Gherab-Martin K and Kalantzis-Cope P (eds) *Emerging Digital Spaces in Contemporary Society: Properties of Technology*. New York: Palgrave Macmillan, 121–122.

Jefferson BJ (2014) Neoliberalizing anticrime activism: Civilian patrols and the postwelfarist citizen. *Spaces and Flows: An International Journal of Urban and ExtraUrban Studies* 4: 59–74.

Jones-Brown D and King-Toler E (2010) The significance of race in contemporary urban policing policy. In Ismaili K (ed) *U.S. Criminal Justice Policy: A Contemporary Reader*. Massachusetts: Jones and Bartlett, 21–48.

Jones-Brown D, Gill J and Trone J (2010) *Stop, Question and Frisk Policing Practices in New York City: A Primer*. New York: Center on Race, Crime, and Justice, John Jay College, City University New York.

Keeling GL and Bratton W (1998) Declining crime rates: Insiders' views of the New York City story. *The Journal of Criminal Law and Criminology* 88: 1217–1232.

Kelling GL and Wilson JQ (1982) Broken windows: The police and neighborhood safety. *The Atlantic* 249: 29–38.

Lefebvre H (1992) *The Production of Space*. Malden, MA: Wiley-Blackwell.

Levine H (2011) *Regarding New York State Senate Bill 5187 (Grisanti): Relating to standardizing penalties associated with marihuana possession and making unlawful possession of small amounts of marihuana a violation punishable by a fine*. Albany: New York Senate.

Loyd J, Mitchelson M, and Burridge A (eds) (2012) *Beyond Walls and Cages: Prisons, Borders, and Global Crisis*. Athens: University of Georgia Press.

Lueck TJ (1992) Study says poverty rose to 25% in New York City. *New York Times*, June 10.

Manning, P (2003) *Policing Contingencies*. Chicago: University of Chicago Press.

Martin L and Mitchelson M (2008) Geographies of detention and imprisonment: Interrogating spatial practices of confinement, discipline, law and state power. *Geography Compass* 3: 459–477.

Milhaud O and Moran D (2013) Penal space and privacy in French and Russian prisons. In Moran D, Gill N, and Conlon D (eds) *Carceral Spaces: Mobility and Agency in Imprisonment and Migration Detention*. Farnham: Ashgate, 167–182.

Moran D (2012) "Doing Time" in carceral space: Timespace and carceral geography. *Geografiska Annaler: Series B, Human Geography* 94: 305–316.

Moran D (2013) Between outside and inside? Prison visiting rooms as liminal carceral spaces. *GeoJournal* 78: 339–351.

Moran D (2014) Leaving behind the "total institution"? Teeth, transcarceral spaces and (re)inscription of the formerly incarcerated body. *Gender, Place and Culture: A Journal of Feminist Geography* 21: 35–51.

Moran D, Gill N, and Conlon D (eds) (2013) *Carceral Spaces: Mobility and Agency in Imprisonment and Migrant Detention.* Farnham, Surrey: Ashgate.

Morin KM (2013) "Security here is not safe": Violence, punishment, and space in the contemporary U.S. penitentiary. *Environment and Planning D: Society and Space* 31: 381–399.

Mountz A (2012) Mapping remote detention: Dis/location through isolation. In Loyd J, Mitchelson M, and Burridge A (eds) *Beyond Walls and Cages: Prisons, Borders, and Global Crisis.* Athens: University of Georgia Press, 91–105.

Mountz A, Coddington K, Catania RT, and Loyd J (2012) Conceptualizing detention: Mobility, containment, bordering and exclusion. *Progress in Human Geography* 37: 522–541.

Nagy A and Podlony J (2008) William Bratton and the NYPD: Crime control through middle management reform. *Yale Case* 7: 1–27.

Nevins J (2012) Policing mobility: Maintaining global apartheid from South Africa to the United States. In Loyd J, Mitchelson M, and Burridge A (eds) *Beyond Walls and Cages: Prisons, Borders, and Global Crisis.* Athens: University of Georgia Press, 19–27.

New York City School–Justice Partnership Task Force (NYSCJPTF) (2013) *Keeping Kids in School and Out of Court: Report and Recommendations.* New York.

New York Civil Liberties Union (NYCLU) (2007). *School to Prison Pipeline Toolkit.* New York.

New York Civil Liberties Union (NYCLU) (2011) *Education Interrupted: The Growing Use of Suspensions in New York City's Public Schools.* New York.

New York Civil Liberties Union (NYCLU) (2014) *Stop-and-Frisk Data.* Available at: www.nyclu.org/content/stop-and-frisk-data.

O'Malley P and Palmer D (1996) Post-Keynesian policing. *Economy and Society* 25: 137–155.

Parenti C (2008) *Lockdown America: Police and Prisons in the Age of Crisis.* London: Verso.

Peck J (2003) Geography and public policy: Mapping the penal state. *Progress in Human Geography* 27: 222–232.

Peck J and Theodore N (2008) Carceral Chicago: Making the ex-offender employability crisis. *International Journal of Urban and Regional Research* 32: 251–281.

Reaves B and Hickman M (2002) *Police Departments in Large Cities, 1990–2000.* Washington, DC: Bureau of Justice Statistics.

Rivera R, Baker A, and Roberts J (2010) A few blocks, 4 Years, 52,000 police stops. *New York Times*, July 11.

Sciarabba A (2009) Community oriented policing and community-based crime reduction programs: An evaluation in New York City. *Professional Issues in Criminal Justice* 4: 27–41.

Shabazz R (2009) "So high you can't get over it, so low you can't get under it": Carceral spatiality and black masculinities in the United States and South Africa. *Souls: A Critical Journal of Black Politics, Culture, and Society* 11: 276–294.

Sibley D and Van Hoven B (2009) The contamination of personal space: Boundary construction in a prison environment. *Area* 41: 198–206.

Siegel L, Perry R, and Gram MH (2006) *Who's Watching? Video Camera Surveillance in New York City and the Need for Public Oversight: A Special Report by the New York Civil Liberties Union.* New York: New York Civil Liberties Union.

Silverman EB (1999) *NYPD Battles Crime: Innovative Strategies in Policing.* Boston, MA: Northeastern University Press.

Smith D (2012) Stop and frisk has lowered crime in other cities. *New York Times*, July 19.

Smith N (1998) Giuliani time: The revanchist 1990s. *Social Text* 57: 1–20.

Smith N (2001) Global social cleansing: Postliberal revanchism and the export of zero tolerance. *Social Justice* 28: 68–74.

Stephan J and Walsh G (2011) *Census of Jail Facilities, 2006*. Washington, DC: U.S. Department of Justice.

United States Bureau of Labor Statistics (2014) Local Area Unemployment Statistics. Available at: http://data.bls.gov/timeseries/LASST360000000000003.

United States Bureau of the Census (1991) *Poverty in the United States: 1990*. Washington, DC: Government Printing Office.

Valentine G and Longstaff B (1998) Doing porridge: Food and social relations in a male prison. *Journal of Material Culture* 3: 131–152.

Van Hoven B and Sibley D (2008) "Just duck": The role of vision in the production of prison spaces. *Environment and Planning D: Society and Space* 26: 1001–1017.

Vera Institute of Justice (2001) *Reinforcing Positive Student Behavior to Prevent School Violence: Enhancing the Role of School Safety Agents*. New York.

Vitale A (2008) *City of Disorder: How the Quality of Life Campaign Transformed New York Politics*. New York: New York University Press.

Wackstein N (1992) Memo to Democrats: Housing won't solve homelessness. *New York Times*, July 12.

Wacquant L (2000) The new "peculiar institution": On the prison as surrogate ghetto. *Theoretical Criminology* 4: 337–389.

Wacquant L (2001) Deadly symbiosis: When ghetto and prison meet and mesh. *Punishment and Society* 3: 95–134.

Wacquant L (2009) *Punishing the Poor: The Neoliberal Government of Social Insecurity.* Durham, NC: Duke University Press.

Wahidin A (2006) A Foucauldian analysis: Experiences of elders in prison. In Powell J and Wahidin A (eds) *Foucault and Aging.* New York: Nova Science Publishers, 115–128.

While A, Jonas A, and Gibbs D (2004) The environment and the entrepreneurial city: Searching for the urban "sustainability fix" in Manchester and Leeds. *International Journal of Urban and Regional Research* 28: 549–569.

Wilson D (2005) *Inventing Black-on-Black Violence: Discourse, Space, and Representation.* New York: Syracuse University Press.

Wilson D (2006) *Cities and Race: America's New Black Ghetto.* London: Routledge.

12 From private to public

Examining the political economy of Wisconsin's private prison experiment

Anne Bonds

Introduction

In 1998, an Edmond, Oklahoma-based private corrections company called Dominion Venture Group, LLC commenced construction on a speculative 1,500-bed prison in Stanley, Wisconsin, a rural community with just over 3,500 residents. Reflecting a common trend in development-driven private prison building, the construction of this facility began without any formal directive or authorization from the state (see also Bonds, 2009). The Stanley prison project coincided with a time of crisis for the Wisconsin Department of Corrections (WDOC): the state's prison population tripled through the decade of the 1990s, swelling to historically unprecedented levels – from 7,117 in 1990 to 20,672 in 2000 – and state facilities were unable to accommodate such growth (Wisconsin Legislative Reference Bureau, 2001). Rather than slowing down, this rate of growth was anticipated to increase. Indeed, construction on the Stanley facility began the same year that the state passed one of the nation's most punitive "truth-in-sentencing"[1] policies, which abolished early release options, instituted extended community supervision, and increased maximum sentences (Greene and Pranis, 2006). "Truth-in-sentencing" lengthened prison sentences and significantly limited the state's parole options, which had previously operated as a mechanism for regulating prison population pressures (Murphy, 2012). Moreover, the state was under increased scrutiny for its incarceration of over 5,000 individuals in the private, Corrections Corporation of America (CCA)-run facilities in Tennessee, Oklahoma, and Minnesota (Wisconsin Legislative Reference Bureau, 2001).

This context created ideal conditions for Dominion's speculative prison venture, which was enthusiastically supported by a coalition of Stanley-area officials and business leaders who – in an all too familiar story – were keen to reap the alleged economic benefits associated with the prison. Despite the fact that state law prevented the WDOC from utilizing in-state private prisons, Dominion's executive director repeatedly expressed confidence that "the state of Wisconsin [would] agree to use the state-of-the-art facility" (*Chippewa Herald*, 1999c). The company further indicated that if the state did not authorize use of the prison, they would house prisoners from outside of the state in the facility (Gunderman and Stetzer, 1999). Stanley area leaders echoed sentiments about the inexorability of the state's use of the facility. As one representative argued, "[t]here's going to be pressure to use the Stanley facility ... [t]hey [the state] virtually have no choice" (*Chippewa Herald*, 1999d).

Ultimately, these postulations were correct. However, the controversial building of the Stanley facility – a project commenced at the local scale by an out-of-state private firm – led to a years-long debate about the role of private contractors in state corrections. In fact, though construction of the facility was completed in April 2000, the building was not put to use by the state until July 2002 (Clement, 2002), and was not funded to operate at full capacity until 2003 (Greene and Pranis, 2006). Importantly, connected to the debate surrounding the Stanley prison was the state's 2001 passage of Wisconsin Act 16. Act 16 was a curious piece of legislation that both authorized the *purchase* of the Stanley prison and also interdicted both the construction of private prison facilities in Wisconsin and the DOC's leasing of any such facilities. The passage of Act 16 raises a number of important questions when considering the historical geography and political economy of prison development. First, what sorts of politics and alliances facilitated and/or stalled the state's purchase of the facility and at what scale? Second, given that the WDOC was incarcerating over 20 percent of the state's prisoners in out-of-state, private CCA facilities, what was the nature of the opposition to the Stanley prison? Finally, did the state's opposition to prison privatization represent a change in state commitments to the logics of incarceration?

Indeed, the prohibition of private prison construction in Wisconsin presents a seemingly optimistic picture, particularly for those concerned with the ethical and political implications of predatory private firms' involvement in practices of incarceration. While Act 16 prevented private prison expansion in Wisconsin, I argue that rather than delimiting prison-building projects in the state, events surrounding the Stanley correctional facility did little to destabilize dramatic prison expansion. Drawing from a range of legislative and newspaper documents, in this chapter I explore the controversies and outcomes of Wisconsin's private prison experiment. I first position the construction of the Stanley prison within Wisconsin's broader carceral expansion. I then chronicle the events surrounding the building of the private prison in Stanley, Wisconsin. Finally, I discuss the ways in which the Stanley prison events reinforced, rather than altered, state politics of incarceration.

A focus on the recent history of private prison development in Wisconsin is important not only to better understand the processes surrounding the siting and placement of correctional facilities and their local impact. It is also significant in documenting a particular historical moment in the state's prison crisis and its punitive turn. An exploration of the dynamics surrounding the Stanley facility reveals the legacy of past corrections priorities and the impacts of state debates about privatization. In examining the political economic dynamics surrounding the construction of the Stanley prison, this chapter charts how historical discourses of privatization shaped Wisconsin's carceral landscape in ways that remain essential in the present: though Dominion is no longer in the incarceration business, the Stanley Correctional Institution currently incarcerates 1,521 inmates.

The Wisconsin prison boom

Mirroring national trends in prison growth driven by the so-called "punitive turn," the number of people incarcerated in Wisconsin rapidly expanded throughout the decade of the 1990s. The well-documented dynamic of mass incarceration in the US is linked to significant ideological, institutional, and political shifts that first emerged in the 1970s, advocating punitive, "tough on crime" corrections policy in favor of more liberal, rehabilitative methods to crime control (Garland, 2001; Western, 2006). This approach, legitimated through the logics of deterrence and incapacitation, criminalized a growing array of minor offenses and significantly increased the prison population. Most notably, the "war on drugs" criminalized low-level drug offenses and attached longer sentences to drug-related crimes, which rapidly expanded imprisonment rates, particularly for people of color, throughout the 1980s and 1990s. By the late 1990s, over 60 percent of those in federal prisons and over 20 percent of state prison populations were incarcerated for drug-related offenses (Murphy, 2012). In addition, state-level mandatory-minimum sentencing policies, "truth in sentencing" reforms, and "three strikes" laws worked to increase prison terms and removed the possibility for parole and sentencing modification. All of these policies worked to send more people to prison and to keep them there for longer.

Wisconsin's prison population grew five-fold in the space of one generation (Green and Pranis, 2006), tripling through the decade of the 1990s alone. In 1980, fewer than 4,000 people were incarcerated, but by the early 2000s the prison population had swelled to nearly 23,000. This expansion rapidly outpaced other states in the region and, indeed, situated Wisconsin as one of the states with the fastest growing prison population in the nation during the 1980s and 1990s (Murphy, 2001; Clement, 2002). This historically unprecedented growth was made possible by the increased incarceration of nonviolent drug offenders facilitated by the state's drug laws, most notably Wisconsin Act 121, passed in 1989, which further extended drug offenses and penalties associated with drug possession and paraphernalia. This policy swept thousands who had little to no prior history into the criminal justice system. In fact, as Greene and Pranis (2006) find in their analysis of the state's prison population, by the mid-2000s nearly half of those in prison for nonviolent drug offenses had no prior felony record. This rapid growth was further reinforced by the state's 1998 "truth in sentencing" policy, noted to be one of the harshest of such policies in the nation. While most states' truth in sentencing policies required that persons incarcerated for serious crimes serve 85 percent of their prison sentence before being eligible for release, Wisconsin's policy required that "*all* prisoners whose offenses were committed after December 30, 1999 … serve *100 percent* of their prison terms behind bars" (Greene and Pranis, 2006: 17).

The impacts of the punitive turn in Wisconsin have been highly racially and geographically uneven. Mirroring national patterns of racialized incarceration, people of color, especially those in the city of Milwaukee, have borne the brunt of the state's failed war on drugs. Black Wisconsinites are incarcerated at a higher rate than any other state (Pawasarat and Quinn, 2013) and are imprisoned at nearly 40 times the rate of White people for nonviolent drug offenses (Greene and Pranis, 2006: 6). A significant portion of the state's prison population incarcerated as a consequence of the state's drug policies was – and continues to be – drawn from Milwaukee County, which contains both the state's large city (Milwaukee) and the largest Black population.

Unsurprisingly, the Wisconsin Department of Corrections was unable to accommodate the soaring prison growth of the 1990s and early 2000s. Following the passage of truth in sentencing in 1998, Republican Governor Tommy Thompson launched an unparalleled prison-building campaign in earnest. From the mid-1990s through the mid-2000s, the state built and acquired nine new prisons – including its infamous "supermax" prison, a facility made up of windowless, concrete cells – and increased the number of beds at existing facilities (Greene and Pranis, 2006). The state's corrections budget quadrupled over the decade of the 1990s and, by the early 2000s, it had reached US$800 million (Clement, 2002). Even with the addition of new facilities and beds, the state was unable to contain its ballooning prison population. Rather than ending its practice of "exporting" prisoners out-of-state, the number of prisoners in private, out-of-state CCA facilities continued to grow. In fact, the number of State of Wisconsin prisoners incarcerated in these facilities accounted for fully one half of *all* prisoners housed in out-of-state prisons nationwide (Greene and Pranis, 2006; Murphy, 2001).

Predatory corrections: privatization and Dominion ventures

As the US prison system expanded throughout the 1980s and 1990s to become the largest in the world, the for-profit corrections industry also grew by leaps and bounds. The pressures wrought by significantly overcrowded prisons created new business opportunities for corrections firms, and claims about the ostensible benefits and cost savings of prison privatization were politically expedient for politicians looking to be both "tough on crime" and fiscally conservative (Greene, 2002; Genter et al., 2013). Although corrections has long been a profitable venture for businesses contracting with the state, the story of the private corrections industry corresponds closely with trends associated with market fundamentalism and neoliberalization (Genter et al., 2013; Greene, 2003). The notion that the private sector could respond more efficiently to the increasing demand for prisons became a common refrain throughout the 1980s, paralleling the rise of market-oriented rhetoric about deregulation, privatization, and fiscal austerity as a means to facilitate cost savings and economic expansion. Meanwhile, private prison firms developed active lobbying campaigns to promote their services, supported by reports and studies from conservative think-tanks like the Heritage Foundation (Greene, 2002). From states' perspectives, prison privatization shifted risk and eliminated the need for large public financing schemes necessary to fund prison construction. Describing this context, Greene (2002: 95) notes that:

Private financing offered prison construction on the installment plan – avoiding bond measures that might require approval by voters, and making end runs around public debt limits. Private construction could cut the red tape entailed in the public procurement process and speed the time to completion. These arguments for privatization were bolstered with generous campaign contributions and political emoluments delivered by squadrons of well-heeled lobbyists.

The first privately run and built corrections facility opened in 1985 in Kentucky and the trend quickly spread, with some 28 other states allowing the operation of private prisons, immigrant detention centers, and other correctional facilities by 1988 (Greene, 2002). Corrections firms grew in scope and scale throughout the 1990s, giving rise to both large and influential corrections firms and growing concerns about the ways in which the bodies of those incarcerated served as capital for powerful corporations. The geography of private prison building during this boom was highly uneven, concentrated primarily in the southern and western regions of the US: Texas built the most private facilities (43), followed next by California (24), then Florida (10), and, finally, Colorado (9) (Murphy, 2001). By 2000, the states with the highest number of persons incarcerated in private facilities were Texas (13,985), followed by Dominion's home state of Oklahoma (6,931). Strangely enough, states *without* private prisons supported the expansion of the for-profit corrections industry in other states by purchasing bed space when they found themselves unable to accommodate their own prison populations. As noted above, Wisconsin was a key player in this dynamic. In 2001, the state incarcerated more that 20 percent of its prison population in private, out-of-state facilities (Murphy, 2001). Describing this dynamic, Murphy (2001) explains that "in 2001, a private prison in Whiteville, Tennessee, a town with 1,100 residents 60 miles east of Memphis, incarcerate[d] as many … as any Wisconsin penal institution."

Although private corrections was represented as a panacea to state prison crises, by the early 2000s the for-profit corrections boom showed signs of slowing down (Murphy, 2001). This decline came amidst a series of reports illustrating the marginal – and often negligible – cost savings associated with private corrections (Genter *et al.*, 2013), myriad scandals in for-profit institutions, and growing concern about profiteering from incarcerated bodies (Greene, 2002; Murphy, 2001). For many, private prisons came to represent the apex of the prison industrial complex, a term first used to describe the many politically and economically vested parties involved in sustaining the practice of incarceration (Davis, 1995). With profit margins in the billions for the two largest corrections corporations (The Corrections Corporation of America [CCA] and the GEO Group), and the embrace of prison privatization as "good politics" (Greene, 2002), concern about for-profit incarceration is appropriate. And yet, it is critically important to note that, even at its peak, the number of people incarcerated in private facilities has always been quite small. In 2000, just 5.8 percent of the incarcerated population in the US was confined in private facilities (Murphy,

2001). Thus, concern about the ethics of predatory corrections firms should not overshadow the fact that the vast majority of the US prison population has always been – and remains – incarcerated in state-run, public facilities. Moreover, the buildup of prison populations that facilitated the rise in for-profit corrections is a consequence of state and federal policy, but not vice versa.

Unlike its larger and more well-known counterparts, such as CCA, Wackenhut, and the GEO Group, much less is known about Dominion's involvement in the private corrections business. Dominion Ventures Group, LLC,[2] the company that built the Stanley prison, is a subsidiary of the Dominion Group of Companies. The Dominion company was founded in 1986 in Edmond, Oklahoma by evangelical entrepreneur Calvin Burgess.[3] Drawing from his background in construction and contracting, Burgess established Dominion with the express goal of "privatiz[ing] governmental functions and projects for which state and federal agencies lacked the funding or flexibility to deliver themselves" (Dominion Group, 2014). Dominion first entered this market through the construction and management of office buildings leased by state and federal governments. The company has built, managed, or leased numerous government facilities, including US Border Patrol stations and several Internal Revenue Service (IRS) and federal office buildings across the states of Kansas, Texas, and New Mexico (Dominion Group, 2014). Although it is difficult to track the precise number of Dominion-built prisons, the company is connected to correctional institutions in Arizona, Colorado, Oklahoma, Minnesota, and Wisconsin, and some of their projects have garnered attention and notoriety. For instance, a Dominion-built and leased prison in McLoud, Oklahoma that was later purchased by the State of Oklahoma came under scrutiny when allegations emerged that prisoners had been denied appropriate drug treatment. In 2009, the company settled a sexual harassment lawsuit with the US Equal Employment Opportunity Commission (EEOC) when it was revealed that male employees were demanding sexual favors from female employees in a Dominion facility in Colorado (EEOC, 2009).

Dominion's murky past certainly reinforces critiques of prison profiteering, but it also raises a number of questions about how the firm came to build a prison in Stanley. In the next section, I discuss the partnership between the City of Stanley and Dominion, and the controversy surrounding the joint venture.

The Stanley prison

Straddling Clark and Chippewa Falls counties in northern Wisconsin, Stanley's historical political economy is based in lumber. The town was founded in 1881, following the arrival of the Wisconsin Central Railway to the area. Soon after, an Eau Claire-based northwestern lumber company established a large lumber mill in Stanley and the activities associated with logging the area's vast pine forests came to dominate the local economy. By the 1920s, after the area's forested land was completely exhausted, the lumber industry significantly declined. Following this collapse, agriculture – especially dairy farming – became increasingly significant to the local economy (City of Stanley, 2014). Although farming remains a significant component of the economy, agricultural activities and the number of farm-related businesses have significantly declined since the 1970s. Over the decade of the 1990s, the city lost 6 percent of its population and one local official claimed that no new housing permits were issued during this same period (Clement, 2002). Compounding this population decline, the area's largest manufacturer laid off more than a quarter of its workforce in the recession of the early 2000s.

As the local economy stagnated, a growth coalition made up of Stanley-area business leaders, city and county officials, and state legislators pursued a prison-led economic development strategy. This agenda was premised on taken-for-granted, though largely unsubstantiated, claims about the economic benefits of prison hosting: namely that prisons provide stable, long-term, well-paid jobs as well as various multiple effects and industrial growth in other sectors like services and retail (Genter *et al.*, 2013; Bonds, 2009, 2013; Gilmore, 2007; Che, 2005; Huling, 2002). Those most vocal and instrumental in facilitating the prison deal – the Stanley Mayor and a Chippewa Falls area state representative – repeatedly invoked such logics to generate local and state support for the prison. The Mayor repeatedly described the prison as being a jobs project that would have a "major economic impact … and bring vitality back to the community" (Clement, 2002: 1). In spite of such common refrains, empirical studies find that economic multipliers of prisons are, in fact, quite limited. For example, prison jobs are often filled by out-of-area employees who commute to prison-hosting cities (Genter *et al.*, 2013). Hooks *et al.* (2004) argue that prisons actually limit future industrial investments and economic development opportunities, and more recent research by Genter *et al.* (2013) finds that private prisons have negative employment impacts in rural communities. All too frequently, prison-hosting communities end up shouldering a range of unanticipated costs associated with the facilities, such as infrastructural updates (Clement, 2002; Bonds, 2013).

While the State of Wisconsin was taking bids, siting, and constructing a number of new public correctional facilities in rural areas across the state, the City of Stanley entered into a joint prison-building venture with the Dominion Ventures. However, Stanley was not the only city in the Dominion considered to site their facility (Gunderman, 2005). The company surveyed a number of small towns in the Chippewa Valley, but ultimately selected Stanley because the city was the only location with adequate infrastructure to support the facility. Stanley's strategy of prison-led development was premised on a direct partnership with Dominion, bypassing state authorization or approval. For its part, Dominion saw the prison as an investment potential, gambling that the state would be hard-pressed not to lease the facility with its corrections facilities operating above capacity and over 20 percent of its prison population housed out of state. Although talks for the prison commenced as early as 1995, construction on the facility did not begin until 1998 (*Chippewa Herald*, 1999f).

As construction was underway, the fact that the project was launched without a state directive became increasingly problematic for state officials and lawmakers unhappy about the speculative venture. Although the facility was backed by some who were eager to take advantage of the bed space that the prison would make available, whether or not the state would lease the facility was still very much in question a year into the prison's construction. Clarifying the state's position, the Attorney General issued a statement in 1999 stating that:

> [p]rivate companies may not operate prisons in Wisconsin or house convicts from other states without permission from the state ... merely constructing a building and calling it an incarceration facility does not in any way mean it can be operated as such.
>
> (*Chippewa Herald*, 1999c)

This sentiment seemed to contradict the fact that the state was sending thousands of prisoners to out-of-state private facilities, a fact that Dominion seized upon to make its case to the public and to the state. A company representative maintained that Dominion was "confident the state of Wisconsin [would] agree to use the state-of-the-art facility ... rather than continue to send thousands of inmates and millions of dollars to for-profit prison in Tennessee, Oklahoma and Texas."

As discussions between the Stanley prison coalition and the state government continued, it became clear that key concerns surrounding the facility centered on staffing and the reservations of the politically influential state employee union, which represented the state's corrections employees: if the state were to lease the facility, would the prison employ private prison employees or would it hire unionized state prison employees? In fact, representatives from the state employees' union made it clear that they were less concerned about the state contracting with out-of-state prisons to temporarily solve overcrowding problems. By contrast, they were extremely uneasy about the long-term labor implications of private prisons in Wisconsin, because they would be permanent (*Chippewa Herald*, 1999b). Ultimately, this problem was resolved when Dominion agreed to staff the facility with state corrections employees.

Even so, the state continued to resist negotiating a lease agreement with Dominion. Much of this opposition remained focused on the speculative nature of the prison and the fact that the facility was built without authorization or approval from the state. One state representative called the process "downright sleazy" (*Chippewa Herald*, 1999f). The Senate's Democratic majority leader, a key figure in the opposition and subsequent fallout, argued that "[i]t is not a good thing for outside developers to come in and decide where to site a prison ... this sort of thing should never happen again" (Gunderman, 2000). Other state representatives, marshaling arguments about the economic benefits of prison hosting, felt that the Stanley–Dominion partnership had foreclosed opportunities for locating the prison in their home districts (Gunderman and Stetzer, 1999). For its part, Dominion attempted to blunt critiques that it had sidestepped state procedures by claiming: "[t]his has always been a project for the state to operate" (*Chippewa Herald*, 1999f). The company further argued that they had sought input from all branches of government since initial plans for the project first emerged, and that they had "found no prohibition for what they had done." However, even as Dominion sought to mollify state mistrust about their intentions, the company was actively seeking contracts from other state agencies and the federal government. At the same time, both the Stanley prison coalition and representatives from Dominion went on a media and lobbying blitz, appealing to lawmakers from both parties to consider the prison project. The Mayor of Stanley made repeated visits to the state capitol to appeal to the Governor and other state officials, pleading the prison case as a jobs issue for Stanley that would also help resolve the state's problem with prison overcrowding (Stetzer, 2000a; Kulish, 2001).

As construction on the prison neared completion, pressure mounted for the state to agree to lease the facility from Dominion. The state assembly – through the leadership of then-Republican State Representative Scott Walker – passed a bill allowing the state to lease private facilities. Walker built support for the provision by emphasizing the DOC's inability to house a significant number of prisoners in state facilities (*Chippewa Herald*, 1999a). Momentum for the facility appeared to be shifting in favor of Dominion and the City of Stanley. The state reached a tentative agreement with the private developer to lease the prison for US$8,093,600 per year (Stetzer, 2000b). However, this momentum came to a dramatic halt when the Stanley prison – and, specifically, the funds necessary to lease the prison – was excluded from the state's 1999 to 2001 budget (Gunderman and Stetzer, 1999). The Senate's Democratic majority leader, Senator Chuck Chvala, refused to move funding toward the prison without setting limits on future private prison construction in the state. He asserted that the state legislature would "not sign a blank check … without assurances that it [private prison building] won't happen again" (*Chippewa Herald*, 2000). Balking at what he characterized as "party politics," the Mayor of Stanley reiterated that local leaders "just wanted the jobs" generated by the prison (Gunderman and Stetzer, 1999).

In April 2000 construction on the Stanley correctional facility was completed. However, with no progress in settling on the terms for a lease, the prison remained empty. Reluctant support for the facility increased from Milwaukee-based representatives who felt it was imperative to bring out-of-state prisoners back to Wisconsin (Kulish, 2001). These sentiments produced an unlikely and provisional alliance between Democratic, urban legislators representing the families of those incarcerated, and conservative, Chippewa Falls-area representatives interested in the prison as a tool of economic development. As one Milwaukee-based official stated, "I don't like the way that the Stanley prison came to be … but it is crazy for Wisconsin to be shipping people out of state when they could be closer to loved ones" (Kulish, 2001). Finally, in October 2001, Wisconsin Act 16 was signed into law, which authorized the purchase of the Stanley Correctional Institution for US$79.9 million, a figure estimated to be 30 percent above the amount that it would have cost the state to build the facility (Snyder, 2009). In addition to authorizing the state's purchase of the facility, Act 16 also prohibited both the building of future private prison facilities in the state and prevented the Department of Corrections from leasing private facilities. Without such a measure, it was unlikely that state democrats would have approved of the purchase. The prison began receiving prisoners in 2002.

Although Act 16 suggested that controversy surrounding the Stanley Correctional Institution was drawing to a close, this was far from the case. In 2002, a legal inquiry revealed that the purchase of the Stanley prison was facilitated by pay-to-play politics rather than an unlikely alliance between anti-private prison Democrats and development-minded Republicans (Walters and Marley, 2006). In fact, a well-known lobbyist testified that State Senate Majority leader Chuck Chvala, who had so adamantly questioned the ethics of Dominion's Stanley prison venture, brokered the purchase of the facility just after he had received US$125,000 in "soft donations" from Dominion Assets. The probe further revealed that employees of Dominion gave smaller donations to the governor (US$4,000) and other state officials from both parties (US$500). Chvala and his wife were convicted of felony corruption charges in February 2002. Dominion evaded criminal charges. However, following the events in Stanley, the company went on record as stating that the Wisconsin prison would be its last speculative prison-building project (Kulish, 2001).

Controversies were reignited in 2009 when, just seven years after the prison began receiving state inmates, a state-commissioned study found that the facility had inadequate security systems and devices. Describing the deficiencies, the report explains: "the prison was constructed by private entity that was not bonded by the current state of Wisconsin construction practices ... which has resulted in the need for extensive remodeling, renovation, and updating of the majority of the facility" (*Chippewa Herald*, 2009). Following the purchase of the facility, the state poured US$20 million into upgrades and repairs (Snyder, 2009). The State Director of Facilities called the project "a white elephant" and another state official argued that "[the state] is so far in the hole with this project, we'll never get out."

From private to public: a cautionary tale

There can be no doubt that the Stanley Correctional Institution was a "boondoggle" of epic proportions for the State of Wisconsin. Dominion Ventures departed Wisconsin – and the speculative prison business – with enormous profits. Stanley area leaders got the prison that they so wanted and claim they were able to leverage other economic development projects from prison-related investments (Gunderman, 2005). And yet, the prison has never employed staff at the levels predicted and a recent empirical study of the employment impacts of private prisons in rural areas points to negative long-term implications (Genter *et al.*, 2013; see also Hooks *et al.*, 2004).

Indeed, Stanley is a cautionary tale. The story exemplifies how predatory corrections firms prey upon the economic desperation of declining rural towns seeking profitable industrial investments. It also illustrates the significant losses accrued by the state as a consequence of corrupt dealings that facilitated a capitulation to Dominion's demands. The tale further demonstrates how geographically contingent scalar politics can confound seemingly straightforward corrections agendas. Gambling on Stanley's desperation and the needs of a budget-strapped state, Dominion Ventures, an extra-local company, commenced the prison-building process, bypassing the state and seeking bids from an array of rural communities in the Chippewa Valley. Both Dominion and the Stanley prison coalition fully expected to easily persuade the state to lease the facility. However, neither group anticipated state-level push-back from officials unwilling to risk the political consequences of upsetting the powerful state corrections union, nor did they anticipate resistance from officials representing other rural areas who saw the facility as a lost opportunity for their own home districts.

Significantly, with the exception of a few vocal Milwaukee area officials, very little of the resistance to the Stanley prison had anything to do with incarcerated populations themselves, the unprecedented growth in state prison populations, or the ethics of for-profit corrections. Although the passage of Act 16 and the prohibition of private, for-profit facilities in the State of Wisconsin points to a potential destabilization of the dramatic growth in incarceration in the state, the event had no impact on prison expansion. In fact, the debate surrounding the Stanley facility never called into question the spectacular growth in state corrections or its larger implications for the welfare of communities. Indeed, incarcerated individuals and their families were largely absent from discussions. Rather, the controversies surrounding the Stanley facility centered on the processes surrounding the siting and staffing of the prison, and these debates were shaped by those who were politically and economically invested in the project. Thus, rather than confounding or problematizing the remarkable growth in the state's prison population, the policies facilitating such growth, or the ethics of for-profit corrections, the debate surrounding the Stanley facility left such questions intact. In so doing, the controversy surrounding the Stanley prison project reinforced the taken-for-granted logics of prison expansion in Wisconsin. In fact, the opening of the Stanley prison did not result in returning home prisoners who were incarcerated out of state, which was a key rationale utilized to bolster support for the Stanley purchase (Stetzer, 2000b). Instead, the facility was utilized to absorb in-state populations from overcrowded facilities.

The building of the Stanley corrections facility represents a coalescing of a particular set of past circumstances and logics: punitive state correctional policy and overcrowding, state budget crises, rural decline, predatory corrections, and taken-for-granted assumptions about privatization. The fixity of prisons means that the outcome of these past priorities and conditions have profound implications for the present. Ultimately, though the prohibition of private prison construction in Wisconsin presents a seemingly optimistic picture, rather than delimiting prison-building projects in the state, the events surrounding the Stanley correctional facility actually worked to reinforce prison expansion. Framed exclusively in the political economic terms of development, building costs, and staffing, the debate eschewed larger questions about policy rationales and the well-being of both urban and rural communities entangled in the dynamics of mass incarceration. The fact that the state was engaged in building nine other public facilities even as the Stanley controversy was unfolding illustrates this fact. Today, the state stands no closer to resolving its prison crisis, with state corrections expenditures now bypassing all spending on higher education. Without a sustained discussion about the logics of incarceration – not just the types of facilities used to serve such a purpose – the state is unlikely to see a shift in this trajectory. Although the politics surrounding private prisons in Wisconsin may have shifted, their spatial legacy remains.

Notes

1 Truth-in-sentencing reforms are associated with the "tough on crime" movements that dramatically reworked state corrections and significantly increased the prison population in the United States throughout the 1980s and 1990s. Key examples of "tough on crime" policies include "three strikes laws," mandatory minimum sentencing for drug offenses, increased sentences for marginal crimes, and truth-in-sentencing policies. Specifically, truth-in-sentencing reforms increased required prison time to be served, and eradicated sentencing modification and early release (Murphy, 2012).
2 Research into Dominion's involvement in private corrections is further compounded by the fact that numerous names are associated with its corrections arm, including Dominion Ventures, Dominion Corrections Services, LLC, and Dominion Asset Services.
3 It is worth noting that Burgess and Dominion have also been involved in a highly criticized agricultural development project in Kenya (Dominion Farms). Although the agribusiness venture is touted as developing food and "advancing the capacity of Kenyans' lives" (www.dominion-farms.com), the company is associated with the degradation and privatization of land and water in the Siaya area in Kenya (Flanders, 2007).

References

Bonds A (2009) Discipline and devolution: Constructions of poverty, race, and criminality in the politics of rural prison development. *Antipode: A Radical Journal of Geography* 41(3): 416–38.
Bonds A (2013) Economic development and relational racialization: "Yes in My Backyard" politics and the racialized reinvention of Madras, Oregon. *Annals of the Association of American Geographers* 103(6): 1389–1405.
Chippewa Herald (1999a) Assembly passes bill allowing private prison leases in Wisconsin. *Chippewa Herald*, November 5. Available at: www.chippewa.com.
Chippewa Herald (1999b) Chief: Contract with Stanley prison possible. *Chippewa Herald*, March 17. Available at: www.chippewa.com.
Chippewa Herald (1999c) Doyle says Stanley prison needs state OK. *Chippewa Herald*, May 29. Available at: www.chippewa.com.
Chippewa Herald (1999d) New prison could have early opening. *Chippewa Herald*, December 27. Available at: www.chippewa.com.
Chippewa Herald (1999e) Senator questions Stanley Prison. *Chippewa Herald*, October 26. Available at: www.chippewa.com.
Chippewa Herald (1999f) Stanley prison owner confident of state support. *Chippewa Herald*, June 6. Available at: www.chippewa.com.
Chippewa Herald (2000a) Chvala sets conditions on prison action. *Chippewa Herald*, February 24. Available at: www.chippewa.com.
Chippewa Herald (2000b) Prison outlook looks brighter. *Chippewa Herald*, December 9. Available at: www.chippewa.com.
Chippewa Herald (2002) Senate Democrats vote to delay opening of Stanley prison. *Chippewa Herald*, April 3. Available at: www.chippewa.com.

Chippewa Herald (2009) Study: Stanley prison security needs upgrade. *Chippewa Herald*, January 9. Available at: www.chippewa.com.

Che D (2005) Constructing a prison in the forest: Conflicts over nature, paradise, and identity. *Annals of the Association of American Geographers* 95(4): 809–831.

City of Stanley (2014) Stanley history. Available at: http://stanleywisconsin.us/history.

Clement D (2002) Big house on the prairie. *Chippewa Herald*, January 19. Available at: www.chippewa.com.

Davis M (1995) Hell factories in the field: A prison industrial complex. *The Nation* 260: 7.

Dominion Group (2010) Home. Available at: www.domgp.com/index.html.

Elbow S (1999) Aggressive firms plan more prisons as state resistance crumbles. *The Capital Times*, October 28. Available at: www.madison.com.

Equal Employment Opportunity Commission (EEOC) (2009) Private prison pays $1.3 million to settle sexual harassment, retaliation claims for class of women. Available at: www.eeoc.gov/eeoc/newsroom/release/10-13-09.cfm.

Flanders L (2007) Obama's ruined homeland. *The Nation*, February 12. Available at: www.thenation.com/article/obamas-ruined-homeland#.

Fourteen communities vie for SuperMax prison (1996) *Beloit Daily News*, September 25. Available at: www.beloitdailynews.com.

Garland D (2001) *The Culture if Control: Crime and Social Order in Contemporary Society*. Chicago, IL: University of Chicago Press.

Genter S, Hooks G, and Mosher C (2013) Prisons, jobs, and privatization: The impacts of prisons on employment growth in rural US counties, 1997–2004. *Social Science Research* 42(3): 596–610.

Gilmore RW (2007) *Golden Gulag: Prisons, Surplus, Crisis, and Opposition in Globalizing California*. Berkeley: University of California Press.

Greene J (2002) Entrepreneurial corrections: Incarceration as a business opportunity. In Mauer M and Chesney-Lind M (eds) *Invisible Punishment: The Collateral Consequences of Mass Imprisonment*. New York: The New Press, 95–113.

Greene J (2003) Lack of correctional services: The adverse effect on human rights. In Coyle A, Neufield R, and Campbell A (eds) *Capitalist Punishment: Prison Privatization and Human Rights*. Atlanta, GA: Clarity Press, 56–66.

Greene J and Pranis K (2006) *Treatment Instead of Prisons: A Roadmap for Sentencing and Correctional Policy in Wisconsin*, prepared by Justice Strategies, New York, and Washington DC. Available at: www.drugpolicy.org/resource/treatment-instead-prisons-roadmap-sentencing-and-correctional-policy-wisconsin.

Gunderman M and Stetzer R (1999) Stanley prison locked out of budget, lawmakers say. *Chippewa Herald*, October 3. Available at: www.chippewa.com.

Gunderman M and Stetzer R (2005) Small towns make the big pitch. *Chippewa Herald*, January 20. Available at: www.chippewa.com.

Hooks G, Mosher C, Rotolo T, and Lobao L (2004) The prison industry: Carceral expansion and employment in U.S. counties, 1969–1994. *Social Science Quarterly* 85: 37–57.

Huling T (2002) Building a prison economy in rural America. In Mauer M and Chesney-Lind M (eds) *Invisible Punishment: The Collateral Consequences of Mass Imprisonment*. New York: The New Press, 197–213.

Kulish N (2001) Homeward bound: States that exported inmates in 1990s have second thoughts now – Wisconsin is bringing back convicts to promote ties to family, generate jobs – "A car, a little gas money." *Wall Street Journal*, December 20, p. A1.

Manta Media Inc. (2013) Dominion Ventures Group L.L.C. Available at: www.manta.com/c/mm2fs9q/dominion-ventures-group-llc.

Mauer M and Chesney-Lind M (2002) *Invisible Punishment: The Collateral Consequences of Mass Imprisonment*. New York: The New Press.

Murphy K (2001) Private prison boom shows signs of slowing. *Stateline Story blog*, The Pew Charitable Trusts, web blog post August 22. Available at: www.pewtrusts.org/en/research-and-analysis/blogs/stateline/2001/08/22/private-prison-boom-shows-signs-of-slowing.

Murphy NM (2012) Dying to be free: An analysis of Wisconsin's restructured Compassionate Release Statute. *Marquette Law Review* 95(4): 1681–1741.

Oklahoma Company Archive (2014) Dominion Correctional Leasing L.L.C. Basic information. Available at: http://oklahoma.company-archive.com/company-profile/dominion-correctional-leasing-l-l-c.4LL.html.

Pawasarat J and Quinn L (2013) *Wisconsin's Mass Incarceration of African American Males: Workforce Challenges for 2013*, prepared by the Employment and Training Institute – University of Wisconsin-Milwaukee, Milwaukee, WI. Available at: www4.uwm.edu/eti/2013/BlackImprisonment.pdf.

Snyder P (2009) Stanley prison sucks up more of Wisconsin's money. *Daily Reporter*, August 18. Available at: www.dailyreporter.com.

Stetzer R (2000a) Prison political game tires leaders. *Chippewa Herald*, February 26. Available at: www.chippewa.com.

Stetzer R (2000b) State locks up Stanley prison lease. *Chippewa Herald*, March 9. Available at: www.chippewa.com.

Stetzer R (2000c) Governor: Feds may lease Stanley prison. *Chippewa Herald*, April 26. Available at: www.chippewa.com.

Stetzer R (2000d) Prison delay tires Stanley residents. *Chippewa Herald*, May 2. Available at: www.chippewa.com.

Stetzer R (2000e) Warden aims for prison opening. *Chippewa Herald*, November 3. Available at: www.chippewa.com.

Walters S and Marley P (2006) To play, you paid, lobbyist reveals: Once secret testimony in Capitol scandal tells how money brought results. *Milwaukee Journal Sentinel*, January 1. Available at: www.jsonline.com.

Western B (2006) *Punishment and Inequality in America*. New York: The Russell Sage Foundation.

Wisconsin Legislative Council (2003) *Legal Memorandum: Drug Laws In Wisconsin: Offenses and Penalties Under Ch. 961, Stats. (The Uniform Controlled Substances Act)*, WLC. Legal Memorandum 2003–2005.

Wisconsin Legislative Reference Bureau (2001) *Budget Briefs: Stanley Prison Purchase*, WLRB, Budget Brief 01–8. October.

13 Afterword

Dominique Moran

Thinking carceral space relationally

Carceral geography has emerged as a vibrant and important subdiscipline of human geography. Although the geographical study of the prison and other confined or closed spaces is relatively new, carceral geography has already made substantial progress, establishing dialogue with cognate disciplines of criminology and prison sociology, and speaking directly to issues of contemporary import such as hyperincarceration and the advance of the punitive state. Carceral geography uses the carceral context as a lens through which to view concepts with wider currency within contemporary and critical human geography, such as mobility, liminality, and embodiment, and now speaks to a diverse audience, with geographical approaches to carceral space being taken up and developed further both within human geography, and the cognate disciplines of criminology and prison sociology (e.g., Crewe *et al.*, 2014).

Carceral geography brings to the study of prisons and imprisonment an understanding of space as relational; dissolving the boundaries between objects and space, and invoking an understanding of objects and processes *as* space, and of space *as* objects and processes, each understandable only in relation to the other in a perpetual process of becoming. The advantage of this approach to space is that, rather than treating it as something that can be classified, delimited, and pigeon-holed, "thinking space relationally" presents a challenge to geographers to consider space as "encountered, performed and fluid" (Jones, 2009: 492). Spaces are seen as "open, discontinuous, relational and internally diverse," and as "complex and unbounded lattice[s] of articulations" (Allen *et al.*, 1998: 143, 65). So, rather than seeing prisons as spatially fixed and bounded containers for people and imprisonment practices, and prison systems as straightforwardly mappable in scale and distance, carceral geography has tended toward an interpretation of prisons as fluid, geographically anchored sites of connections and relations, both connected to each other and articulated with wider social processes through and via mobile and embodied practices (Moran, 2015). This approach has delivered, within carceral geography scholarship, a focus on experience, performance, and mutability of prison space; revealed the porosity of the prison boundary and mobilities within and between institutions; and shown a concern for the ways in which meanings and significations are manifest within fluid and ever-becoming carceral landscapes.

Carceral geography has considerable scope to contribute to the advancement of geographical scholarship, bringing a perspective on "closed" spaces that has much to offer critical human geography, both in relation to a new "territory" for exploration, and to regime shifts in the carceral landscape. The key challenge it faces, while contributing to contemporary discourse within human geography, and remaining open to transdisciplinary engagement, is to retain a critical perspective that helps bring about progressive social transformation.

Carcerality's usable past

This volume addresses this challenge head-on, through the notion of a "usable past," interpreted diversely and individually by the contributors, who were asked to consider, in their scholarship, the ways in which their historical research could inform and influence issues of contemporary import. We opened this book with recourse to Tosh's (2008: 22–23) notion of a "critical applied historical geography" and of the usable past, and this interwoven thread of the "critical" and the "applied" has been drawn through each subsequent chapter. Our authors have outlined the ways in which their scholarship can be interpreted and deployed in advocating for progressive social transformation, in terms of making the past relevant, purposeful, and usable.

In the three themed parts of the book, the authors have brought their own different and innovative interpretations to bear on the notion of the usable past, and its relevance and purposefulness. In Part I, where the focus is on historical geographies and the techniques of incarceration, including insights from prisoner experiences, **Hemsworth** urges geographers to be attuned to soundscapes, bringing to light previously overlooked characteristics of carceral spaces, and thus deepening and empathetically enhancing both our understandings of spaces of incarceration, and of the experiences of their occupants. **Story** uses her genealogical study of solitary confinement to remind us that, rather than deleting prisoners from the sphere of citizenship, and recalling the conditions of the camp after Agamben, a "usable past" of the solitary confinement unit is in fact a living history of struggle, of political agency, and of social being. By retelling the history of solitary confinement in a way that highlights prisoners' active resistance to the oppressive structures of the prison regime, her "usable past" not only carries the potential to recuperate prisoners' political power, but also reminds us that even the most extreme technologies of state violence *can* be challenged.

Shabazz's chapter diagnoses the ways in which carceral power in the form of the segregation and targeted policing of Chicago's South Side neighborhoods gave rise to the Black gangs of the 1960s and shaped the shifting, circulating gender performances of Black men both in Illinois prisons and local neighbourhoods. Tracing this usable past in the present, he reflects directly on the lessons for these neighborhoods from this genealogy, in terms of scaled-back policing, investment in communities, and a reversal of what he calls the march toward mass incarceration.

In Part II, which details cultural-historical practices and understandings related to the decommissioning or re-commissioning of correctional institutions, **Turner** and **Peters'** chapter performs a dual function, documenting the ways in which a prison museum itself seeks to make the past "usable" for its visitors while also engaging with these performances of the past in order to better understand how imprisonment is understood and engaged with in the present. The ways in which the usable past of their prison museum both relies on, and reinforces, public perceptions of imprisonment generated in media constructions and in the public imagination, speaks to the notion of "public punitiveness" (Garland, 2001; Greer and Jewkes, 2005; Young, 2003; Frost, 2010); where harsher sentencing policy is often attributed in part to demands for "punishment" from what is assumed to be an increasingly punitive public, whose views therefore indirectly and directly affect criminal justice policy.

Walby and **Piché** take a different perspective on the prison museum, exploring the repurposing of small, decommissioned carceral sites as local history museums, and focusing on the role of local historical and heritage societies in curation work that allows tourists to enjoy fleeting encounters with carceral architecture, space, relics, narratives, and experiences. Their usable past highlights the ways in which the retasking of these small sites makes accessible *some* aspects of Canada's penal history but obscures others, and most notably the development of large penal institutions in Canada, through which larger regional facilities have replaced these smaller jails. Refashioned by local historical societies in ways that promote heritage appreciation, these museums do little to articulate the usable past of imprisonment with contemporary practices of imprisonment and punishment. **Medlicott**'s chapter bucks the trend of current scholarship in carceral retasking, in that she takes as her focus conversion *to* a purpose of incarceration, rather than a re-purposing *from* a prison into something else. She argues that the layout and structure of Shaker sites rendered them "usable" to other total institutions, making them literally a "usable past." More subtly, though, she also teases out the ways in which the adaptive reuse of sites from the needs of the arguably self-incarcerated Shakers, to those of a formally institutionalized or incarcerated collective demonstrates that the "usable past" is present in both the figurative and the most literal sense. Finally, **Draper** explores the potential for collective prison art projects to become a form of memory building, connecting the struggle for the re-creation of memory of the past with the practice of signification of the present. She emphasizes a notion of a usable past that reclaims the act of remembering and historicizing carceral space for those who have themselves been imprisoned.

In Part III on political-economical analyses of the landscape of prison development, expansion, and contraction, **Anderson *et al.***'s geographically and temporally expansive chapter re-conceptualizes penal transportation as a coherent if locally divergent penal and labor regime underpinned by the dynamics of imperial space across a variety of global contexts. Here, a usable past can emerge through a subtle framing of the relationship between the past and the present, one that allows for a more systematic integration of historical research into

carceral geography, through the integrated study of broad connections and specific contexts which respects their irreducibly context- and time-bound nature. In
the specific context of rural New York State, **Norton** finds a usable carceral past
in the expansion of the penal estate as a spatial fix to economic and social crises,
and points out that focusing on carceral infrastructure spurs us to consider what
else could have been done with the evident social capacity for organization and
development that found itself channeled into the local enabling of mass incarceration. By contrast, focusing on New York City, **Jefferson** details the rise of
hyperpolicing, a distinct mode of law enforcement rendering targeted areas more
"prisonlike," and argues that the usability of this carceral past is in part in its
potential to anticipate prospective modes of carceral governance in the light of
post-recession austerity measures. Finally, **Bonds'** chapter argues that rather
than limiting prison-building projects in Wisconsin, the prohibition of private
prison construction actually worked to reinforce prison expansion. Narrowly
focused on development, building costs, and staffing, local debate eschewed
larger questions about policy rationales and the well-being of communities in
mass incarceration. A critical and applied perspective here is evident in the
observation that without sustained discussion about the *logics* of incarceration –
and not just the types of facilities to be built – there is unlikely to be a shift in
policy.

In taking a critical and applied perspective to their various contexts and case
studies, the contributing authors have traced usable pasts which address not only
the lived experience of incarceration, the geographies of carceral systems, and
the relationship between the carceral system and the state, but have maintained a
steady focus on the locus of power which shapes and reshapes the system itself,
and which molds punitive opinion in ways that serve to sustain incarceration.
The notions of the usable past expounded here are as diverse as the temporal and
spatial nature of the case studies. Some authors demonstrated the potential of
specific types of methodology to the integration of historical materials with contemporary concerns, while at the same time avoiding a "presentist" perspective.
Others detailed the conscious construction of intentionally usable, physical pasts
in the form of penal museums, the repurposing of prisons, or indeed what might
be termed the "pre-purposing" of specific sites later to be used for incarceration.
In each case, subtle arguments are teased out about the ways in which these
practices reflect and reinforce a punitive carceral imagination. In addition,
turning attention to really-existing prison systems, their siting, construction, and
operation at various historical junctures, others have pragmatically considered
what may be usefully learned from these experiences, and have additionally considered the ways in which these pasts might prove most "usable" in anticipating
the operation of post-austerity governance.

The work of these authors, from diverse disciplinary backgrounds, and in
divergent contexts, moves both carceral geography and historical geography on
in important and innovative ways. Broadening the scope of historical geographical scholarship of prison spaces, these chapters also provide temporal depth and
richness to carceral geography. Coalescing at a critical moment in contemporary

penal practice, with the expansion of "workfare" and "prisonfare" policies, the criminalization of immigration, and the expansion of the carceral estate, carceral geography can sometimes seem so focused on the pressing concerns of the present that it risks neglecting the lessons of the past. Through a shared desire to make the past relevant, to demonstrate its purpose, and, critically, to *make a difference*, the contributors to this volume offer a compelling case for the usable carceral past.

References cited

Allen J, Massey D, and Cochrane A (1998) *Re-thinking the Region.* London: Routledge.

Crewe B, Warr J, Bennett P, and Smith A (2014) The emotional geography of prison life. *Theoretical Criminology* 18(1): 56–74.

Frost NA (2010) Beyond public opinion polls: Punitive public sentiment and criminal justice policy. *Sociology Compass* 4(3): 156–168.

Garland D (2001) *The Culture of Control.* Oxford: Oxford University Press.

Greer C and Jewkes Y (2005) Extremes of otherness: Media images of social exclusion. *Social Justice* 32(1): 20–31.

Jones M (2009) Phase space: Geography, relational thinking, and beyond. *Progress in Human Geography* 33(4): 487–506.

Moran D (2015) *Carceral Geography: Spaces and Practices of Incarceration.* Farnham: Ashgate.

Tosh J (2008) *Why History Matters.* Basingstoke: Palgrave Macmillan.

Young J (2003) Searching for a new criminology of everyday life: A review of the "culture of control." *British Journal of Criminology* 43(1): 228–243.

Index

Page numbers in **bold** denote figures.

Milton Keynes UK
Ingram Content Group UK Ltd.
UKHW040104071024
449327UK00019B/806